Low Thermal Expansion Glass Ceramics

Schott Series on Glass and Glass Ceramics
Science, Technology, and Applications

Low Thermal Expansion Glass Ceramics
ISBN 3-540-24111-6

Fibre Optics and Glass Integrated Optics
ISBN 3-540-58595-8

The Properties of Optical Glass
ISBN 3-540-58357-2

Thin Films on Glass
ISBN 3-540-58597-4

Electrochemistry of Glasses and Glass Melts, Including Glass Electrodes
ISBN 3-540-58608-3

Surface Analysis of Glasses and Glass Ceramics, and Coatings
ISBN 3-540-58609-1

Analysis of the Composition and Structure of Glass and Glass Ceramics
ISBN 3-540-58610-5

Mathematical Simulation in Glass Technology
ISBN 3-540-43204-3

Hans Bach
Dieter Krause
Editors

Low Thermal Expansion Glass Ceramics

Second Edition

With 156 Figures
and 21 Tables

Springer

Editors

Dr. Hans Bach
Prof. Dr. Dieter Krause
Schott AG
Hattenbergstraße 10
D-55122 Mainz, Germany

Library of Congress Control Number: 2004116340

ISBN-10 3-540-24111-6 Springer Berlin Heidelberg New York
ISBN-13 978-3-540-24111-9 Springer Berlin Heidelberg New York
ISBN 3-540-58598-2 1st ed. Springer-Verlag Berlin Heidelberg New York

Springer is a part of Springer Science+Business Media

springeronline.com

Printed in Germany

Typesetting: Computer to film from editors data
Production: LE-TEX Jelonek, Schmidt & Vöckler GbR, Leipzig

Printed on acid-free paper 56/3141/YL 5 4 3 2 1 0

Foreword

This book, entitled *Low Thermal Expansion Glass Ceramics*, is one of a series reporting on research and development activities on products and processes conducted by the Schott AG.

The scientifically founded development of new products and technical processes has traditionally been of vital importance at Schott and has always been performed on a scale determined by the prospects for application of our special glasses. The scale has increased enormously since the reconstruction of the Schott Glaswerke in Mainz. The range of expert knowledge required for that could never have been supplied by Schott alone. It is also a tradition in our company to cultivate collaboration with customers, universities, and research institutes. Publications in numerous technical journals, which since 1969 we have edited to a regular timeplan as *Forschungsberichte* – 'research reports' – formed the basis of this cooperation. They contain up-to-date information on various topics for the expert but are not suited as survey material for those whose standpoint is more remote.

This is the point where we would like to place our series, to stimulate the exchange of thoughts, so that we can consider from different points of view the possibilities offered by those incredibly versatile materials, glass and glass ceramics. We would like to show scientists and engineers, interested customers, and friends and employees of our firm the knowledge that has been won through our research and development at Schott in cooperation with the users of our materials.

The results documented in the volumes of the Schott Series are of course oriented to the tasks and targets of a company. We believe it will become quite clear that here readers can nevertheless – or rather for that reason – find demanding challenges for applied research, the development of process engineering, and the characterization of measurement practice. Besides realizability, the profitability of solutions to customers' problems always plays a decisive role.

The first comprehensive presentation of research findings after the reconstruction of the factory in Mainz was edited by Prof. Dr. Dr. h.c. Erich Schott in 1959. It was entitled *Beiträge zur angewandten Glasforschung* – 'contributions to applied glass research' (Wissenschaftliche Verlagsgesellschaft m.b.H., Stuttgart 1959). Since then, there has been an extraordinary worldwide increase in the application of glass and glass ceramic materials. Glass fibers and components manufactured from them for use in lighting and traffic engineering or in telecommunications, high-purity and highly homogeneous glasses for masks and projection lenses in electronics, or glass ceramics with zero expansion in astronomy and in household appliance technology are only some examples. In many of these fields Schott has made essential contributions.

Due to the breadth and complexity of the field in which Schott is active, many volumes are needed to describe the company's research and development results. Otherwise it would be impossible to do full justice to the results of fundamental research work and technological development needed for product development. Furthermore, it is necessary to give an appropriate description of the methods of measurement and analysis needed for the development and manufacture of new products.

Apart from *Low Thermal Expansion Glass Ceramics*, five volumes, entitled *The Properties of Optical Glass*, *Thin Films on Glass*, *Analysis of the Composition and Structure of Glass and Glass Ceramics*, *Electrochemistry of Glasses and Glass Melts, Including Glass Electrodes*, and *Mathematical Simulation in Glass Technology* have already been published. Another two volumes, entitled *Surface Analysis of Glasses and Glass Ceramics, and Coatings* and *Fibre Optics and Glass Integrated Optics*, are in preparation and will be published in the next few years. Glasses for various applications in industry and science and their properties are being considered, and melting and processing technologies described.

With the presentation – in part detailed – of the work required for the development of successful products, Schott employees are giving all their interested colleagues who work in the field of science and technology an insight into the special experiences and successes in material science, material development, and the application of materials at Schott. Contributions from scientists and engineers who work in university and other research institutes and who played an essential role in Schott developments complete the survey of what has been achieved. At the same time such results show the need for the collaboration mentioned above.

In all the volumes of the series the fundamental issues from chemistry, physics, and engineering are dealt with, or at least works are cited that enable or assist the reader to work his or her way into the topics treated. We see this as indispensable because, with the series, Schott has a further goal in view. We aim to provide all future business partners from branches of industry where glasses and glass ceramics have not been applied so far with knowledge they can use in cooperation with Schott. Furthermore, the series may serve to fill gaps between the basic knowledge imparted by material science and the product descriptions published by Schott. Those who have already done business with our company may find the survey of fundamentals useful in extending collaboration to further business areas.

To make each volume sufficiently intelligible, the necessary fundamentals from chemistry, physics, and engineering are described or referred to via citations. We see this as the best way to enable all our potential business partners who are not already familiar with glass and glass ceramics to compare these materials with alternatives on a thoroughly scientific basis. We hope that this will lead to intensive technical discussions and collaborations on new fields of applications of our materials and products, to our mutual advantage.

Every volume of the Schott Series will begin with a chapter providing a general idea of the current problems, results, and trends relating to the subjects treated. These introductory chapters and the reviews of the basic principles are intended to be useful for all those who are dealing for the first time with the special properties of glass and glass ceramic materials and their surface treatment in engineering, science, and education.

Many of our German clients are accustomed to reading scientific and technical publications in English, and most of our foreign customers have a better knowledge of English than of the German language. It was, therefore, mandatory to publish the Schott Series in English.

The publication of the Schott Series has been substantially supported by Springer. We would like to express our special thanks to Dr. H.K.V. Lotsch and Dr. H.J. Kölsch for advice and assistance in this project.

The investment of resources by Schott and its employees to produce the Schott Series is, as already stated, necessary for the interdisciplinary dialogue and collaboration that are traditional at Schott. A model we still find exemplary today of a fruitful dialogue between fundamental research, glass research, and glass manufacture was achieved in the collaboration of Ernst Abbe, Otto Schott, and Carl Zeiss. It resulted in the manufacture of optical microscopes that realized in practice the maximum theoretically achievable resolution. It was especially such experiences that shaped the formulation of the founding statute of the Carl Zeiss Foundation, and the initiative for the Schott Series is in accord with the commitment expressed in the founding statute "to promote methodical scientific studies".

Mainz, March 2005

Dieter Krause
Vice President R & D (retd.)

Preface to the Second Edition

The second edition has been corrected and supplemented. Several additions have been made wherever it was necessary, with major changes in chapters 3 and 4. For their valuable advice and support in updating these chapters we are indebted to Dr. Peter Naß and Dr. Peter Hartmann.

In 2004, the Schott Group became an incorporated company (Schott AG). Former company names such as “Schott Glas” or “Jenaer Glaswerk” are sometimes used in this book for historical reasons.

We thank the authors for reading, correcting and updating their contributions, Mrs. Karin Langner-Bahmann for processing all the figures, and Mrs. Wiltrud Witan for revising the English. We also thank Springer for supporting this edition.

March 2005 Hans Bach, Dieter Krause

Preface

The main aim of the Schott Series volume *Low Thermal Expansion Glass Ceramics* is to describe research and development necessary to produce glass ceramics having low thermal expansion coefficients and to present some products manufactured at Schott, which are the results of a successful development. The book is conceived as a monograph. However, the individual chapters have been written by different or several authors, who are themselves active in the corresponding fields of research and development. Thus the reader is given direct access to the experience of these authors.

To give the reader a view of the extraordinary material "glass ceramic", the volume opens with a general survey of the development of glass ceramics and their important fields of application and the aims, limits, and the current state of new developments.

Schott has significantly contributed to the development and production technology of glass ceramics during the last four decades. The subsequent chapters treat in detail the scientific basis of glass ceramics, the special properties of glass ceramics to reach outstanding functionality in use, and the technology designed for the economic production of technical equipment at Schott. Results from two fields of application are presented where research and development have been particularly successful: from household appliances and from equipment for optics and astronomy. This presentation necessarily also includes a rough description of production methods and machines, whose design has been dictated by the processing parameters derived from basic research.

To obtain a basis for a deeper understanding of the problems encountered in the development and production of glass ceramics so that they can be considered as engineered materials, the reader is introduced in the first section of the second chapter to the special field of crystal chemistry and physics of high-quartz and keatite-type aluminosilicates. In this section it is explained why useful properties might be obtained based on certain types of solid solutions of these silicates. The development of a variety of those solid solutions appears to be possible, whose coefficient of thermal expansion and grain size distributions can be adapted to applications. Products consisting of these silicates can only be shaped economically if the forming methods of the conventional glass production are applicable prior to crystallization.

Further investigations are, therefore, necessary to decide whether this is possible or not. In the second and third sections of the second chapter, methods are described which allow us to determine the basic parameters for the production of a glass ceramic and the development of a glass ceramic based on lithium-alumino-silicate solid solution crystals.

The subsequent chapters, 3 and 4, are devoted to the description of the development and application of glass ceramics for household appliances and for optical instruments.

Chapter 3 reports on the special research and development that forms the basis of the production of the glass ceramic Ceran®. This glass ceramic has meanwhile become well-known worldwide since it is widely used for cooktops. It is also described how Ceran® is able to meet the requirements for functionality and appealing appearance in the kitchen.

The properties of other glass ceramic products have also been tailored to special household applications: the properties of the glass ceramic Robax® were adapted to its use as stove windows. Chemical strengthening of the surface of another glass ceramic used for cooktops can improve their functionality.

Chapter 4 is dedicated to the development and application of the glass ceramic Zerodur®. Several applications in optics are possible due to the unique properties of this material. The production of pieces made of this material for optical instruments with large dimensions has successfully been performed at Schott. In particular, pieces having very large dimensions (as they are used for very large telescopes) can be manufactured at Schott. The reader is informed about technologies and basic research and may well imagine that plenty of scientific and technological knowledge had to be acquired until the production of such materials and, particularly, casting and forming the products of large dimensions could be controlled. Chapter 4 closes with illustrations of the use of Zerodur® for special optical instruments and for mirrors with large dimensions for astronomy.

In Chaps. 3 and 4 the technologies are also described, which had to be adapted to the parameters to make upscaling of large dimensions possible in production. The finally chosen technologies for forming, nucleation, and the thermal treatment during nucleation and crystal growth guarantee both reproducibility of the required properties of the glass ceramics and the most economic production possible.

The properties of the glass ceramics and their varieties are also reported on. Additionally, methods of quality assurance are mentioned, which are necessary to grant the mechanical, thermal, and chemical properties and the demanded final shapes of the products. The considerable effort in the analysis of bulk material and surface analysis, which must be applied in basic research and development to study the appropriate parameters for nucleation and crystal growth, could not be covered by the present book. The reader is referred to the two volumes on analysis and surface analysis to appear in this series.

The results given in Chaps. 2, 3, and 4 inform the reader about how the findings of basic research determine the processing of glass ceramics. A close cooperation between scientists and engineers is imperative in developing the special technologies and suitable equipment and ensuring the most economic reproduction of the required properties of the different glass ceramics and glass ceramic components designed for different applications. Thus this volume contributes to filling the gap of knowledge about engineering which exists between the published results on the basics of glass ceramics and the catalogue data on glass ceramics provided by producers. The form of the presentation of both the results on the basics and the technology can, moreover, be useful for teaching.

In summary, all the information given in the present book exemplifies the successful transfer of results from basic science reported on in Chap. 2 into products and production processes via a fruitful cooperation between research, development, and technology, and, last but not least, our customers.

I would like to thank all the authors of this book for their steady and pleasing cooperation.

I have received further valuable help from many colleagues. For critical reading of the manuscript I thank in particular Dr. Hartmut Höness, Dipl.-Phys. Alfred Jacobsen, Dipl.-Phys. Hans Morian, Dr. Rudolf Müller, Dr. Peter Naß, Dr. Wolfgang Pannhorst, Dipl.-Ing. Norbert Reisert, Dr. Erich W. Rodek, and Dipl.-Ing. Hinnerk Schildt.

For their advice and help and converting technical drawings into figures appropriate for publication my thanks go to Dipl.-Ing. Heinrich Nilgens and Dipl.-Ing. Wolfgang Walch.

Additionally, I am indebted to several employees of Springer-Verlag, especially to Barbara S. Hellbarth-Busch and to Peter Straßer, production-editor, for helping us to overcome the difficulties involved in producing manuscripts ready for printing. I am thankful to Dr. Victoria Wicks for copy-editing this volume and Andy Ross for various advice. For their help in solving text processing problems I am indebted to Frank Holzwarth, also of Springer-Verlag and to Kurt Mattes, Heidelberg.

I am very grateful to Dipl.-Math. Sieglinde Quast-Stein, Schott Glaswerke, who, with her knowledge and experience provided substantial support in the implementation of the software guidelines supplied by Springer-Verlag.

I also thank Dipl.-Grafik-Designer Werner Paritschke, Mainz, for the creation of the numerous computer graphics needed to illustrate the texts.

I would especially like to thank Mrs. Angela Gamp-Paritschke, M. A., Schott Glaswerke, for translations from German into English, for the corrections of manuscripts submitted in English, and for her enthusiasm in performing all the hard work necessary to prepare manuscripts ready for printing.

September 1995 Hans Bach

Contents

1. Overview

Wolfgang Pannhorst

1.1 The Invention of Low Expansion Glass Ceramics

Glass ceramics are the result of two independent lines of research activities in the USA in the 1940s and 1950s which, when combined, opened up the view of a family of materials with a high potential for new applications. One route of research was performed at Corning Glass Works by Stookey who investigated the nucleation of glasses. While for a long time his research centered around photonucleation of opal and colored glasses with crystalline phase contents of less than 5%, he one day found accidentally that some of these photonucleated glasses can be transformed by an annealing process to highly crystalline materials with a very fine microstructure, i.e., with crystal sizes in the range of microns. In a further research effort he found that similar results may be obtained by using special additives, so-called nucleating agents, instead of the photonucleation process. His fundamental patent [1.1] discloses that TiO_2 acts as such a nucleating agent in a rather large number of glass systems.

The other route started with the discovery by *Hummel* in 1951 [1.2] that crystalline aggregates of β-eucryptite (Li_2O-Al_2O_3-$2SiO_2$) display a negative volume expansion. People immediately realized that this observation opens up the perspective of developing materials without any expansion in some temperature intervals, thus creating thermoshock resistant or dimensionally highly stable materials. As a consequence, an intensive research activity started to find out whether this observation is restricted to β-eucryptite alone or whether a whole family of materials can be defined, which in the following will be called high-quartz solid solution (h-quartz s.s.) crystals. Although at the beginning of these activities the intention was to produce sintered ceramics, the main field of research interest very quickly switched over to the development of glass ceramics when it became apparent that the Li_2O-Al_2O_3-SiO_2 materials family also belongs to those glass ceramic systems that can be nucleated very efficiently by TiO_2. The glass ceramic approach has two major advantages over the ceramic approach: (a) very fine-grained microstructures can be produced; (b) high-speed glass manufacturing processes can be used. The latter advantage is certainly off-set to some extent by the so-called ceramization process, an annealing process by which the original glass is transformed into the glass ceramic.

1.2 Basic Research

Since about 1960 many glass companies as well as glass research institutions have started research in the field of glass ceramics; their work mainly centered on the Li_2O-Al_2O_3-SiO_2 system (LAS). The investigations within the LAS system were directed into three areas: (a) solid solution formation in the h-quartz structure; (b) improvement of the efficiency of the nucleating agents; (c) stability field of the h-quartz s.s. crystals.

In the area of h-quartz s.s. formation the main results were as follows. The β-eucryptite composition is a special, stoichiometric one within a whole family of solid solution crystals which all can be derived from the h-quartz (SiO_2) crystal structure. Substituting Si^{4+} in the quartz structure by Al^{3+} may be achieved over a wide percentage range when charge compensation is admitted by either Li^+ [1.3–5], Mg^{2+} [1.6], or Zn^{2+} [1.5, 7]. While quartz shows a reversible phase transition at 573 °C from low to high quartz, the h-quartz structure is stable at room temperature when roughly more than 20 mol% of the SiO_2 is substituted by one of the pairs (Al_2O_3, Li_2O), (Al_2O_3, MgO), or (Al_2O_3, ZnO) [1.5]. These three coupled substitutions are possible up to approximately 50 wt% replacement of SiO_2. Finally, it was found [1.8, 9], within the substitutional field of 20–50 wt% of SiO_2 by one of the coupled pairs, that up to about 70 wt% of the remaining SiO_2 may be replaced by $AlPO_4$, still with the h-quartz s.s. crystal structure being the metastable phase which crystallizes first from glasses and which does not undergo any high–low transition when being cooled to room temperature.

Although these substitutions principally widen the field of chemical compositions, thus allowing not only optimization of the coefficient of thermal expansion (CTE) but also other important properties, the range of useful compositions is decreased by the fact that the substitutions influence the thermal expansion characteristics. Generally speaking, the $LiAlO_2$ substitution results in a strongly negative, the $ZnAl_2O_4$ substitution in a slightly negative, and the $MgAl_2O_4$ substitution in a strongly positive CTE, whereas the $AlPO_4$ substitution has only a small effect on the CTE.

Stookey discovered that TiO_2 acts as a very efficient nucleating agent in LAS-based glass ceramics, whereas *Tashiro* and *Wada* [1.10] found that ZrO_2 additions have a similar effect. Finally, *Sack* and *Scheidler* [1.11] showed that the utilization of both nucleating oxides has advantages, especially by lowering the temperature of the transformation of the base glass into the glass ceramic.

It would be desirable that the h-quartz s.s. phase with its excellent thermal expansion characteristics is stable up to high temperatures so that the material may be used in high temperature applications. Unfortunately, this is not the case with compositions that show the most promising property combinations for applications and whose main components lie in the field Li_2O-Al_2O_3-$nSiO_2$ with $5 < n < 7$. The h-quartz s.s. crystalline phase is a metastable phase which transforms into the keatite s.s. phase (for explanation see next paragraph) at temperatures between 800 and 950 °C depending on the time–temperature conditions. For applications in which service temperatures of 700 °C or more are to be expected during the life span of the product the choice of the compositions is constrained by the careful observation that the transformation of h-quartz s.s. to keatite s.s. occurs at high enough temperatures.

Keatite is the name of an SiO_2 modification which does not occur in nature but can be synthesized under hydrothermal conditions. As in quartz, solid solution formation is also possible in keatite. Well documented is the solid solution formation in the system Li_2O-Al_2O_3-SiO_2 (LAS), especially along the line Li_2O-Al_2O_3-$nSiO_2$, with n ranging from 4 to 10 [1.12]. The composition with $n = 4$, i.e., Li_2O-Al_2O_3-$4SiO_2$, is called β-spodumene, and in many papers this composition is seen as the starting point of the solid solution formation so that the phases which in this book will be called keatite s.s. phases are often also called β-spodumene s.s. phases.

The keatite s.s. phases in the LAS system are also very interesting phases in that respect as they show negative or only small positive thermal expansion characteristics. They may, therefore, also serve as materials with low expansion. Solid solution formation for keatite has not been investigated as systematically as that for h-quartz, probably because there are indications that the solid solution formation is much more restricted for keatite than for h-quartz. This information has been derived from the investigation of the phase transformation of some of the low expansion materials based on the h-quartz s.s. phase. During these transformations often formation of spinels ($MgAl_2O_4$ or $ZnAl_2O_4$) [1.13, 14] or cordierite ($2MgO$-$2Al_2O_3$-$5SiO_2$) [1.13, 15] is observed, indicating that the solid solution formation with ZnO or MgO replacing Li_2O is rather limited. Nevertheless, low-expansion glass ceramics based on keatite s.s. phases are of interest when either high service temperatures up to approximately 1100 °C or increased strength are important application requirements in addition to the low-expansion characteristics.

The development of the low-expansion glass ceramics is a commercially very successful part of a much broader effort to understand nucleation and crystal growth phenomena, on the one hand, and to develop products based on the glass ceramic approach on the other hand. These fields have, therefore, been the topics of many conferences and the accompanying proceeding volumes [1.16–19] as well as of several books [1.20–23].

1.3 Main Fields of Application

Product development mainly centered around the three product ideas cookware, range tops for kitchen stoves, and telescope mirror blanks.

Since the development of the heat-resistant borosilicate kitchen ware, glass makers have longed to develop cookware which may be used in all household situations, i.e., which can be stored in the refrigerator, placed onto the hot stove directly from the refrigerator, and used as attractive dishes on the table. With the low-expansion glass ceramic this vision could become reality. There were even two possible solutions. Based on the keatite glass ceramic a version which resembles porcelain was possible, while the h-quartz glass ceramics offered the possibility to produce a tinted, transparent variant, similar to the borosilicate kitchen ware. Both variants have been developed and launched to the market by Corning Inc. as well as by Schott. The two Corning brands, Pyroceram® for the white, opaque keatite glass ceramic and Vision® for the transparent h-quartz glass ceramic are still being sold, but both Schott products, Ceradur® (keatite glass ceramic) and Jena 2000® (h-quartz glass ceramic) have been withdrawn from the market after some time. Although the cookware withstands all situations where either high thermal gradients or thermal shock occur, the glass ceramics have one important deficiency: their low thermal conductivity compared to metals. When the stove does not heat the bottom of a pot or a pan uniformly, hot spots form locally and burn the food rather easily; the situation is improved by coating the outer bottom side with a metallic layer to increase thermal conductivity, but nevertheless the food is burned in a glass ceramic pot more easily than in a metallic one. So the versatility to use one attractive glass ceramic pot in all household situations is probably offset by this disadvantage; this is assumed as glass ceramic cookware was only able to seize a minor portion of the whole cookware market within more than 30 years.

The application of glass ceramics as range tops for kitchen stoves requires very similar material properties as in the case with cookware. Again the most obvious requirement is to chose a material that withstands steep thermal gradients with temperature differences of 500 °C and more. The low-expansion glass ceramics seem to be very well suited for such an application due to their low CTE values of 0.1×10^{-6}/K or even less for the temperature interval 20–700 °C. Compared to electrically heated kitchen stoves with local steel heating plates a glass ceramic solution offers the advantage that the range top is made from one flat piece without elevated heating zones and gaps between heating zone and the remaining range top area; so pots will not tilt when placed improperly onto the heating zone and food that has fallen onto the range top can easily be removed. One of the possible weaknesses of a glass ceramic as a range top, which worried the material developers, was the strength of the material. Average strength values for newly delivered samples (as-received samples) lie around 150 MPa for the keatite glass ceramics and 100 MPa for the h-quartz glass ceramics.

Also for this product idea, developments based on the keatite glass ceramic as well as on the h-quartz glass ceramic were performed and launched to the market. Corning Inc. and PPG developed white, opaque variants based on keatite glass ceramics while Schott, and later Corning together with Saint Gobain, developed a strongly tinted, partly transparent variant, based on h-quartz glass ceramics. It is the glass ceramic Ceran® developed by Schott which has been very successful and continuously seized a larger part of the market for electrically heated kitchen stoves. Currently, the application of glass ceramic range tops is extended to kitchen stoves which use as energy supply either gas or gas and electricity in one and the same stove.

The requirements for telescope mirror blanks are quite different; here dimensional stability of the shape of the blanks with variations in the temperature is of prime importance. The temperature interval considered is much smaller than in cookware or range tops and mainly encompasses temperature differences of 50–150 °C. Within these temperature intervals the CTE has to be as close to zero as possible. Although the precision optical glass ceramic Zerodur® from Schott is now used in a variety of applications, the development was driven by astronomers looking for a mirror blank material with a lower CTE than that for fused silica, i.e., lower than 0.5×10^{-6}/K. When such developments were performed in the 1960s at Owens Illinois and at Schott, the CTE target value was $0\pm0.15\times10^{-6}$/K. One of the reasons why astronomers were attracted by the new low expansion glass ceramics has to do with their experience with large glass mirror blanks made of borosilicate glass. For a material to be used in large mirror blanks an important requirement is that the large blanks can be produced with high homogeneity; as astronomers had gained good experiences with large glass castings they hoped that similar results would be obtained with glass ceramics, because the first production step of a glass ceramic is identical to normal glass production. While this assumption was true for the Schott glass ceramic Zerodur® it probably was less true for the glass ceramic Cer-Vit® from Owens Illinois; the production of this latter glass ceramic, which had good property characteristics, has later been abandoned, probably due to quality problems.

The development of the glass ceramic Zerodur® was stimulated by a request of the Max-Planck-Gesellschaft in 1966 to produce eleven mirror blanks of different sizes, with the largest being 3.6 m in diameter and 0.6 m thick. In the 1970s, telescope mirror blanks were the main application for precision optical glass ceramics; only slowly new applications were found which required their unique materials properties. The two most important applications are laser gyroscopes for navigation purposes and mirrors for reflective optical systems in chip lithography.

Rapid developments also took place in two main directions in the design of telescope mirrors, which became lighter and larger. Schott participated in several development programmes investigating different approaches with respect to their feasibility. The outcome of these studies revealed that the preferred solutions are thin menisci which are supported by active actuators. These developments are highlighted by the present engagement of Schott in the production of thin menisci of more than 8 m in diameter and with 29 mm thickness.

1.4 Current Developments

The incentives for new developments in the area of low-expansion glass ceramics are rather low. It seems that the basic understanding of these materials has been achieved so that ideas for further improvements mainly address optimizations of the production processes and of existing products. These ideas are well-kept secrets inside each company and are not communicated to the scientific community.

An area which still lacks a good understanding of all the phenomena observed is the irradiation of low-expansion glass ceramics with high-energy particles in space. Although the h-quartz s.s. crystals are destroyed very rapidly when irradiated with 100 keV electrons in the electron microscope, they withstand irradiation with 0.3–1.5 MeV electrons in space or in space simulation experiments. This seems to be due to the larger areas irradiated in the latter experiments [1.24]. Nevertheless, the glass ceramic is compacted when exposed to space irradiation [1.25] and the compaction seems to be higher in the simulation experiments than in space experiments [1.26]. This discrepancy has not been understood. On the other hand, a good understanding of the compaction of the Zerodur® glass ceramic in space is of great importance for space antennae designers. Their preferred material is Zerodur® because of its high dimensional stability with temperature variations. It is of special interest to them whether this high-dimensional stability is reduced by space radiation or not.

Around 1970, tough and strong glasses were developed by reinforcing glass matrices with carbon fibers [1.27–29]. In ambient atmosphere the maximal service temperature of these interesting materials lies in the range of 400–450 °C due to the low oxidation resistance of the carbon fibers. When the Nicalon SiC fibers [1.30] were launched on the market the development of fiber-reinforced glasses received a new push. As the Nicalon fibers are stable in oxidizing conditions up to about 1200 °C, fiber-reinforced glass ceramics were rapidly invented.

The LAS glass ceramic was tested as one of the first matrices for fiber-reinforced glass ceramics [1.31–33]. This matrix seemed to be very attractive because of its low expansion, thus giving the opportunity to develop strong and tough composites with low thermal expansion up to about 1100 °C. In the meantime, the effort in the development of these materials has been reduced considerably because of two drawbacks. One drawback is observed in most, perhaps all, fiber-reinforced glass ceramics with Nicalon SiC fibers. The high toughness of these composites is strongly related to a thin (100 nm) carbon layer which forms during processing at the interface between the fiber and the matrix [1.34–36]. This layer starts to burn off in oxidizing atmosphere at about 800 °C, turning the tough material into a brittle one [1.33, 35, 37]; so the temperature stability does not exceed 800 °C as was originally hoped. The second drawback is related to the properties of the LAS matrix. The Li ions in the glass ceramic are very mobile and thus can be exchanged very easily for other monovalent ions. As H^+ is present in many technical environments the material is not stable under these conditions; thus, the material is not suited for many of the applications originally considered.

In recent years LAS glass ceramics received some new attention by the observation that keatite s.s. [1.38] and h-quartz s.s. [1.39] containing glass ceramics can be formed by the photonucleation process. But it still has to be shown that low-expansion glass ceramics can be obtained via the photonucleation process.

As will be described in Sect. 1.5, glass ceramics with a very fine-grained microstructure can also be obtained by powder processing of glasses. This processing sequence has also been applied to the base glass of the glass ceramic Zerodur®, resulting in a glass ceramic which has the main property characteristics of Zerodur® with only a few exceptions [1.40]. The most important exception is the lack of transparency; the Zerodur® variant processed via powder processing is white opaque. The inspection of this variant for internal quality is more complicated than is the case with normal Zerodur®. On the other hand, the key property of Zerodur®, its low CTE, is reproduced for the powder variant as easily as for normal Zerodur®. Powder processing allows the production of bodies with very complex shapes which are not amenable to glass forming. It was this advantage of powder processed bodies which initiated the development of a powder variant of Zerodur®; but up to now no economically attractive product could be identified.

1.5 Other Glass Ceramics

Because of the unique property achievable with low-expansion glass ceramics this material class attracted most attention at the beginning of the development of glass ceramics; but of course many researchers tried to apply the basic ideas of the formation of glass ceramics to other composition fields, and often they were very successful. Nowadays there is no principal reason why the glass ceramic approach could not work in other composition fields, although the specific details have to be worked out for each field separately.

The economic success of the development of glass ceramics depends not only on successful materials development, but even more on the requirement that the glass ceramic is expected to significantly outperform all competing materials by at least one property for the following reasons. Already the production of glasses is a relatively expensive process. The melting at high temperatures is capital and energy intensive. This deficiency can be compensated for when either a highly automated process of high speed can be used to produce mass products or when material properties such as transparency or homogeneity are of prime importance. For the production of glass ceramics the base glasses have to be converted into glass ceramics by an additional heat treatment process for which the speed is usually lower than that for the production of the glass articles; so this process not only increases capital and energy costs but also slows down the production speed. Altogether, even the mass production of glass ceramics is an expensive process which will be able to compete with glasses, ceramics, metals, or plastics only if the performance of these articles is much superior to any competing product. When the demand for a product is rather low and mass-production processes cannot be applied, the costs of a glass ceramic component will even be higher compared with a component produced from one of the less expensive materials; in these situations the benefit of a glass ceramic solution has to be even more pronounced than for a mass-demand product.

Only a few examples of the many successful developments in the field of glass ceramic materials, which have been performed over the last 35 years, are mentioned here, further information can be found in [1.16–23, 57].

The first glass ceramic developed is based on a photonucleation process [1.20, 41]. In this case, Ag_2O is added to the glass composition in the Li_2O-SiO_2 base system in small amounts. By the photonucleation process the Ag ions are converted to atoms which first agglomerate and then precipitate as tiny Ag crystals; these crystals act as nucleation sites for the precipitation of the main crystal phase of the glass ceramic, Li_2SiO_3. The glass ceramic has the surprising property that it is leached by diluted hydrofluoric acid by a factor of 20 faster than the base glass. Using lithographic methods, very fine-structured parts can be fabricated which have, for example, been used in ink-jet printers [1.42].

So-called machinable glass ceramics reveal another outstanding property. Their main crystal phase forms micas or other plate-like crystals which are easily cleavable. When pieces of these glass ceramics are machined with conventional metal-working tools they do not break into pieces as normal glasses typically do, but they can be machined easily to the desired shape. The machinable glass ceramics have this ability because the cracks which are created during the machining process do not run catastrophically through the whole piece but are deviated at the small plate like crystals and, at the same time, split into several others so that the energy which is introduced into the working piece is absorbed by the formation of many small cracks. Machinable glass ceramics can be applied in very different areas. One area is the prototyping of components for new equipment or systems in those cases in which the fabrication of the few pieces needed from the material of optimal choice is too expensive at that time [1.43]. Other applications concern medical areas, for example, dental restoration [1.44, 45] or bone restoration [1.46]. For these applications the original idea to produce a machinable glass ceramic has been further extended to materials which are at the same time biocompatible or bioactive.

Biocompatibility and bioactivity are outstanding properties on their own. Several glass ceramics have been developed which show high biocompatibility or bioactivity but which are not machinable with conventional metal-working tools [1.47].

The idea to produce ceramic-like materials with a fine microstructure by controlled devitrification of base glasses was soon extended to procedures other than the controlled volume nucleation and crystallization of base glasses. Relatively fine-grained glass ceramics can also be obtained by sintering and crystallization of glass powders to dense bodies.

For their development it is important to know that glass grains nucleate rather easily from the grain surface, so that several or many nuclei are formed at the surface of each grain. This seems to be true for the original grains even after they have coalesced to larger grains [1.48, 49]. To produce fine-grained microstructures in dense, sintered glass ceramics it is, therefore, necessary to use fine-grained powders and to control the crystallization process so that densification proceeds crystallization. The fabrication of fine-grained powders with submicron grain sizes is easily achieved with modern powder fabrication techniques. The main attention nowadays is, therefore, concentrated on the goal of achieving high densities before crystallization starts and hampers any further densification.

The production of sintered glass ceramics was proposed in 1965 [1.50]. Two important glass ceramic products are produced nowadays by this procedure: panels for walls of buildings and multilayer substrates for silicon chips. Panels for walls were successfully developed in the 1970s by NEG in Japan under the tradename Neoparies® [1.51]. The panels outperform equivalent ones from natural rocks such as marble or granite and still meet the price range acceptable for architects. The development of cordierite multilayer substrates for mainframe computers by IBM since the second half of the 1970s has been a very convincing example of how a special feature can only be achieved by glass ceramic processing [1.52, 53]. Co-firing of about 30 layers with conducting pastes between them is a prerequisite to multilayer substrate fabrication. To replace Al_2O_3 multilayer substrates with Mo wires which are co-fired at about 1500 °C, a materials combination was sought, which could be fired below 950 °C, so that highly conductive metals (Ag, Cu) could be used, and for which the dielectric constant of the layer material is approximately 5 compared with 9.4 for Al_2O_3. Starting with a cordierite base glass, which during firing (sintering) transforms into the crystalline form, allowed all the requirements to be met.

The powder processing route has also been used in the fabrication of bioactive glass ceramics [1.54] mentioned in Sect. 1.4. A more recent development combines powder processing of glass ceramics and sol-gel techniques. Although producing the glass powder by the sol-gel technique instead of glass melting is a straightforward approach, seeding the sol prior to gelation with seed crystals [1.55–57] will probably open up new avenues to control the development of the microstructure of a base glass.

References

1.1 S.D. Stookey: "Method of making ceramics and product thereof", US Patent 2 920 971 (January 12, 1960)
1.2 F.A. Hummel: "Thermal expansion properties of some synthetic lithia minerals", J. Am. Ceram. Soc. **34**, 235–239 (1951)
1.3 H. Saalfeld: "Struktur und Ausdehnungsverhalten von Li- Al-Silikaten", Ber. Dt. Keram. Ges. **38**, 281–286 (1961)
1.4 A.G. Alekseev, M.V. Zasolotskaya: "Certain cystalline phases separating in glasses of the system Li_2O-Al_2O_3-SiO_2 with small amounts of TiO_2", in *The Structure of Glass*, Vol. 7, ed. by E.A. Porai-Koshits (New York, Consultants Bureau 1966) pp. 172–175
1.5 J. Petzoldt: "Metastabile Mischkristalle mit Quarzstruktur mit Oxidsystem Li_2O-MgO-ZnO-Al_2O_3-SiO_2", Glastechn. Ber. **40**, 385–396 (1967)
1.6 W. Schreyer, J.F. Schairer: "Metastable solid solutions with quartz-type structures on the join SiO_2-$MgAl_2O_4$", Z. Kristallogr. **116**, 60–82 (1961)
1.7 G.H. Beall, B.R. Karstetter, H.L. Rittler: "Crystallization and chemical strengthening of stuffed β-quartz glass-ceramics", J. Am. Ceram. Soc. **50**, 181–190 (1967)
1.8 J. Petzoldt: "Der Einbau von P_2O_5 in metastabile Mischkristalle mit Quarzstruktur des Grundsystems Li_2O-MgO-ZnO-Al_2O_3-SiO_2", Glastechn. Ber. **41**, 181–189 (1968)
1.9 A.J. Perotta, R.O. Savage: "Beta-eucryptite crystalline solutions involving P^{5+}", J. Am. Ceram. Soc. **50**, 112 (1967)
1.10 M. Tashiro, M. Wada: "Glass ceramics catalyzed with zirconia", in *Advances in Glass Technology*, part 2 (Plenum, New York 1963) pp. 18–19
1.11 W. Sack, H. Scheidler: "Einfluß der Keimbildner TiO_2 und ZrO_2 auf die sich ausscheidenden Kristallphasen bei der Bildung von Glaskeramik", Glastechn. Ber. **39**, 126–130 (1966)
1.12 W. Ostertag, G.R. Fischer, J.P. Williams: "Thermal expansion of synthetic β-spodume and β-spodume-silica solid solutions", J. Am. Ceram. Soc. **51**, 651–654 (1968)
1.13 M.D. Karkhanavala, F.A. Hummel: "Reactions in the system Li_2O-MgO-Al_2O_3-SiO_2: I, the cordierite–spodumene join", J. Am. Ceram. Soc. **36**, 393–397 (1953)
1.14 Q.Q. Chen, P.L. Gai, G.W. Groves: "Microstructure and grain growth in Li_2O-Al_2O_3-SiO_2 glass ceramics", J. Mat. Sci. **17**, 2671–2676 (1982)
1.15 S. Ray: "Solid solutions in the keatite crystal lattice", J. Am. Ceram. Soc. **54**, 213–215 (1971)
1.16 Symposium on Nucleation and Crystallisation in Glasses and Melts, (Am. Ceram. Soc., Columbus, Ohio 1962)
1.17 L.L. Hench, S.W. Freiman (Eds.): *Advances in Nucleation and Crystallisation in Glasses* (Am. Ceram. Soc., Columbus, Ohio 1971)
1.18 J.H. Simmons, D.R. Uhlmann, G.H. Beall (Eds.): "Nucleation and crystallisation in glasses", in *Advances in Ceramics*, Vol. 4 (Am. Ceram. Soc., Columbus, Ohio 1982)
1.19 M. Weinberg (Ed.): *Nucleation and Crystallisation in Liquids and Glasses*, Ceramic Transact., Vol. 30 (Am. Ceram. Soc., Westerville, Ohio 1993)
1.20 P.W. MacMillan: *Glass Ceramics*, 2nd ed. (Academic Press, London 1979)
1.21 A.I. Berezhnoi: *Glass Ceramics and Photo-Sitalls* (Plenum, New York 1970)
1.22 Z. Strnad: *Glass-Ceramic Materials* (Elsevier, Amsterdam 1986)
1.23 M.M. Lewis: *Glasses and Glass-Ceramics* (Chapman and Hall, London 1989)

1.24 S. Thiel: "Modelle zur Beschreibung elektronenmikroskopischer Strahleffekte an Gläsern und Glaskeramiken", PhD Thesis (Halle 1994)
1.25 P.L. Highby et al.: "Radiation effects on the physical properties of low-expansion-coefficient glasses and ceramics", J. Am. Ceram. Soc. **71**, 796–802 (1988)
1.26 W. Pannhorst: "Response of the glass ceramic Zerodur® to simulated and actual space radiation", Proc. XVII Int. Cong. Glass, Beijing, Oct. 1995
1.27 R.A. Sambell, D. Bowen, D.C. Phillips: "Carbon fiber composites with ceramic and glass matrices, part 1, discontinuous fibers", J. Mat. Sci. **7**, 663–675 (1972)
1.28 R.A. Sambell, A. Briggs, D.C. Phillips, D. Bowen: "Carbon fiber composites with ceramic and glass matrices, part 2, continuous fibers", J. Mat. Sci. **7**, 676–681 (1972)
1.29 D.C. Phillips, R.A. Sambell, D.H. Bowen: "The mechanical properties of carbon fiber reinforced Pyrex", J. Mat. Sci. **7**, 1454–1464 (1972)
1.30 S. Yajima, K. Okamura, J. Hayashi, M. Omori: "Synthesis of continuous SiC fibres with high tensile strength", J. Am. Ceram. Soc. **58**, 324–327 (1976)
1.31 J.J. Brennan, K.M. Prewo: "Silicon carbide fibre reinforced glass-ceramic matrix composites exhibiting high strength and toughness", J. Mat. Sci. **17**, 2371–2383 (1982)
1.32 K.M. Prewo: "Tension and flexural strength of silicon carbide fibre reinforced glass ceramics", J. Mat. Sci. **21**, 3590–3600 (1986)
1.33 K.M. Prewo, J.J. Brennan, G.K. Layden: "Fiber reinforced glasses and glass ceramics for high performance applications", Am. Ceram. Soc. Bull. **65**, 305–313, 322 (1986)
1.34 J.J. Brennan: "Interfacial characteristics of glass-ceramic matrix/SiC fiber composites", J. Phys. **49**, C5-791–809 (1988)
1.35 R.F. Cooper, K. Chyung: "Structure and chemistry of fiber-matrix interfaces in silicon carbide fiber-reinforced glass-ceramic composites: An electron microscopy study", J. Mat. Sci. **22**, 3148–3160 (1987)
1.36 R. Chaim, A.H. Heuer: "The interface between (Nicalon) SiC fibres and a glass ceramic matrix", Adv. Ceram. Mat. **2**, 154–158 (1987)
1.37 E. Bischoff, M. Rühle, O. Sbaizero, A.G. Evans: "Microstructural studies of the interfacial zone of a Sic-fiber-reinforced lithium aluminium silicate glass-ceramic", J. Am. Ceram. Soc. **72**, 741–745 (1989)
1.38 J. Sehgal: "Mechanism and kinetics of photonucleation and strengthening of glass by light induced crystallisation", PhD Thesis (Alfred University, NY 1992)
1.39 H.M. Hoffmann: "Zur Photosensibilität und Photokristallisierbarkeit von Gläsern, aufbauend auf dem System Li_2O-Al_2O_3-SiO_2", PhD Thesis (Mainz)
1.40 W. Pannhorst, W. Semar: "Sintered Zerodur bodies with complex shapes", in *Proc. 4th Int. Symp. New Glass*, New Glass Forum, Tokyo, 127–128 (1993)
1.41 S.D. Stookey: "Photosensitively opacifiable glass", US Patent 2 684 911 (July 27, 1954)
1.42 G.H. Beall: "Design of glass ceramics", Rev. Sol. State Sci. **3**, 333–354 (1989)
1.43 D.G. Grossman: "Machining a machinable glass ceramic", Am. Machinist, 139–142 (May 1978)
1.44 K.A. Malament, D.G. Grossman: "The cast glass-ceramic restoration", J. Prosthetic Dent. **57**, 674–683 (1987)
1.45 G.H. Beall: "Glass-ceramics: recent developments and applications", in *Nucleation and Crystallisation in Liquids and Glasses*, Ceramic Transact., Vol. 30, ed. by M. Weinberg (Am. Ceram. Soc., Westerville, Ohio 1993) pp. 241–266
1.46 W. Vogel, W. Höland: "The development of bioglass ceramics for medical applications", Angew. Chem. Int. Ed. **26**, 527–44 (1987)

1.47 T. Kokubo: "Novel bioactive materials derived from glasses", Bol. Soc. Esp. Ceram. VID. **31C**, Vol. 1, 119–138 (1992)
1.48 F.B. Siebers, N. Greulich, W. Kiefer: "Manufacture, properties and application of open-pore sintered glasses and open-pore sintered glass-ceramics", Glastechn. Ber. **62**, 62–73 (1989)
1.49 T. Rudolph, D.V. Szabo, W. Pannhorst, K.-L. Weisskopf, G. Petzow: "Microstructural development of a cordierite glass ceramic during sintering, Part 1: Microscopic experiments", Glastechn. Ber. **64**, 218–224 (1991)
1.50 W. Sack: "Glas, Glaskeramik und Sinterglaskeramik", Chem.-Ing.-Techn. **37**, 1154–1165 (1965)
1.51 M. Wada, S. Kawamura: "Some glass ceramics for special applications", Bull. Inst. Chem. Res. Kyoto Univ. **59**, 256–265 (1981)
1.52 R.R. Tummala: "Ceramic and glass-ceramic packaging in the 1990s", J. Am. Ceram. Soc. **74**, 895–908 (1991)
1.53 J.U. Knickerbocker: "Overview of the glass-ceramic/copper substrate – a high-performance multilayer package for the 1990s", Am. Ceram. Soc. Bull. **71**, 1393–1401 (1992)
1.54 T. Kokubo: "Glass ceramic A-W: Structure, properties and biomedical applications", Proc. 5th Int. Otto Schott Colloquium, Glastechn. Ber. Glass Sci. Technol. **67C**, 105–110 (1994)
1.55 J.C. Debsikdar, O.S. Sowemimo: "Effect of zirconia addition on crystallinity, hardness, and microstructure of gel-derived barium aluminosilicate, $BaAl_2Si_2O_8$", J. Mat. Sci. **27**, 5320–5324(1992)
1.56 M. Chen, P.F. James, W.E. Lee: "Synthesis of monoclinic celsian from seeded alkoxide gels", J. Sol-Gel Sci. Techn. **1**, 99–111 (1994)
1.57 P.F. James: "Glass ceramics: new compositions and uses", J. Non-Cryst. Solids **181**, 1–15 (1995)

2. The Scientific Basis

2.1 Structure, Composition, Stability, and Thermal Expansion of High-Quartz and Keatite-Type Alumino-Silicates

Gerd Müller

2.1.1 Crystal Structures

With very few exceptions the numerous polymorphs of SiO_2 all consist of SiO_4 tetrahedra linked through their corners, thus forming three-dimensional framework structures. The topology of the tetrahedral linkage and the efficiency of space filling are different for the polymorphs. For a given type of framework, for example, that of quartz or cristobalite, space filling can be improved by so-called displacive transformations from a more open high-temperature form (e.g., "high", "h", or "β-quartz") to a denser form stable at lower temperatures ("low" or "α-quartz"). These transformations do not change the topology of the framework, i.e., chemical bonds in a crystal can be deformed, but are not broken and rearranged.

Many alumino-silicates of the general formula $MAlSi_xO_{2x+2}$, M being a univalent ion or one half of a bivalent ion, also have framework structures. There, AlO_4 tetrahedra are linked up with SiO_4 tetrahedra in an ordered or disordered way, with the M ions occupying cavities in the framework and providing charge neutrality. Some alumino-silicates adopt the frameworks of SiO_2 polymorphs; M.J. Buerger coined the term "stuffed derivatives" for them.

The high-quartz and keatite-type alumino-silicates are such stuffed derivatives. Both types of framework are relatively dense, compared with, for example, those of tridymite and cristobalite. Therefore, only small univalent or divalent M ions can be accommodated in their cavities, Li ions being typical. Indeed, solid solutions of the composition $LiAlSi_xO_{2x+2}$ exist with both types of framework over wide ranges of x.

Figures 2.1 and 2.2 show projections of the two types of structures for the composition of $LiAlSi_2O_6$. The AlO_4 and SiO_4 tetrahedra are disordered in both cases. For convenience, they are called "TO_4" tetrahedra, T standing for a statistical mixture of $Al_{1/3}$ $Si_{2/3}$ in the case considered here.

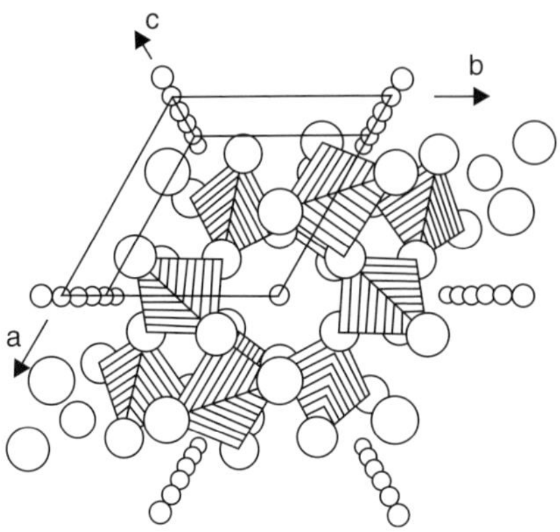

Fig. 2.1. The crystal structure of h-quartz-type $LiAlSi_2O_6$ viewed along [001]. Small circles: Li positions. Large circles: O. Si in centers of tetrahedra

The high-quartz-type structure consists of 6- and 8-membered rings of TO_4 tetrahedra. It has hexagonal symmetry, space group $P6_222$ (or $P6_422$). The tetrahedra form helical chains about the 6_2-screw axis which, in Fig. 2.1, goes through the origin of the unit cell. Channels parallel to the c axis are thus

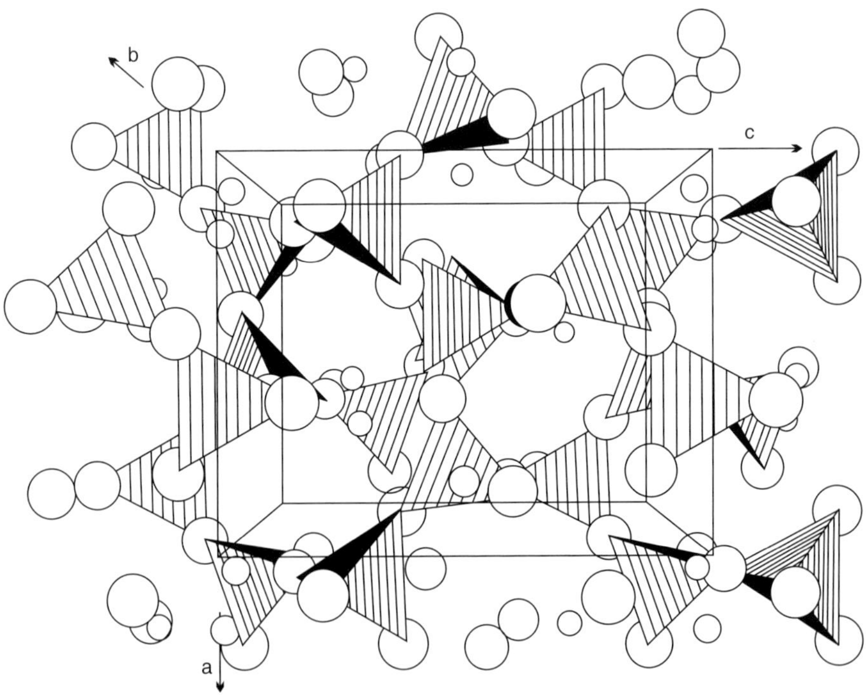

Fig. 2.2. The structure of keatite-type $LiAlSi_2O_6$ viewed along [010]. Symbols as in Fig. 2.1

formed; the Li ions occupy positions with tetrahedral oxygen coordination in these channels [2.1].

β-eucryptite, $LiAlSiO_4$, has a structure of the same topology, but normally with an ordered distribution of the AlO_4 and SiO_4 tetrahedra, which leads to a doubling of the c lattice parameter. The Al/Si order, in turn, causes rather complicated, temperature-dependent ordering of the Li ions in the channels which, at temperatures below about 400 °C, also requires doubling of the a parameter. Because of these complications and because of its interest as a superionic conductor, β-eucryptite has been the object of many X-ray and neutron diffraction studies ([2.2–7] and further references cited there).

In contrast to the high-quartz structure, the keatite structure is tetragonal, space group $P4_32_12$ (or $P4_12_12$) and consists of 5- and 7-membered rings of TO_4 tetrahedra. The 5-rings can be seen in Fig. 2.2, they are connected to adjacent 5-rings by common T–O–T bridges, thus forming ribbons of 5-rings extending in the a and b directions. There again, the Li ions occupy cavity positions with tetrahedral oxygen coordination. These occur pairwise, only one per pair is occupied statistically [2.8].

In spite of the differences in both symmetry and ring sizes there is a rather close relationship between the high-quartz and keatite structures: instead of the hexagonal unit cell of high-quartz with edges a_1, a_2, and c, another one can be defined by $a'_1 = a_1 + a_2 + c$; $a'_2 = a_1 + a_2 - c$; $c' = a_1 - a_2$. This new, larger unit cell is pseudotetragonal and comes quite close to the unit cell of the keatite structure both in size and content, see Fig. 2.3 [2.9, 10]. However, a transition between the two phases is not displacive in character, but requires bond breakage and reconstruction.

Besides the tetrahedral positions occupied by the Li ions, both structures contain sites with a distorted octahedral oxygen coordination. In the high-quartz structure these sites are located in the channels, in the mid position between adjacent tetrahedral sites. In the keatite structure the octahedral sites are also halfway between the pairs shown in Fig. 2.2. In isostructural alumino-silicates containing cations other than Li^+ it has been found that Mg^{2+} occupies the octahedral positions in the high-quartz-type structure [2.11], whereas Zn^{2+} prefers the tetrahedral sites [2.12]. Consequences of these site preferences will be discussed below.

2.1.2 Compositional Ranges and Stability

In the general formula of the alumino-silicates, $MAlSi_xO_{2x+2}$, both the type of cations M and the parameter x can vary considerably. Furthermore, the SiO_2 content can be replaced by $AlPO_4$ to some extent. Finally, limited substitution of Al in the aluminate component $MAlO_2$ by other small trivalent or even bivalent atoms has also been reported.

Because of their relatively poor space filling, the high-quartz-type and keatite-type alumino-silicates have ranges of thermodynamic stability only at low pressure and elevated temperature. Large compositional areas are not

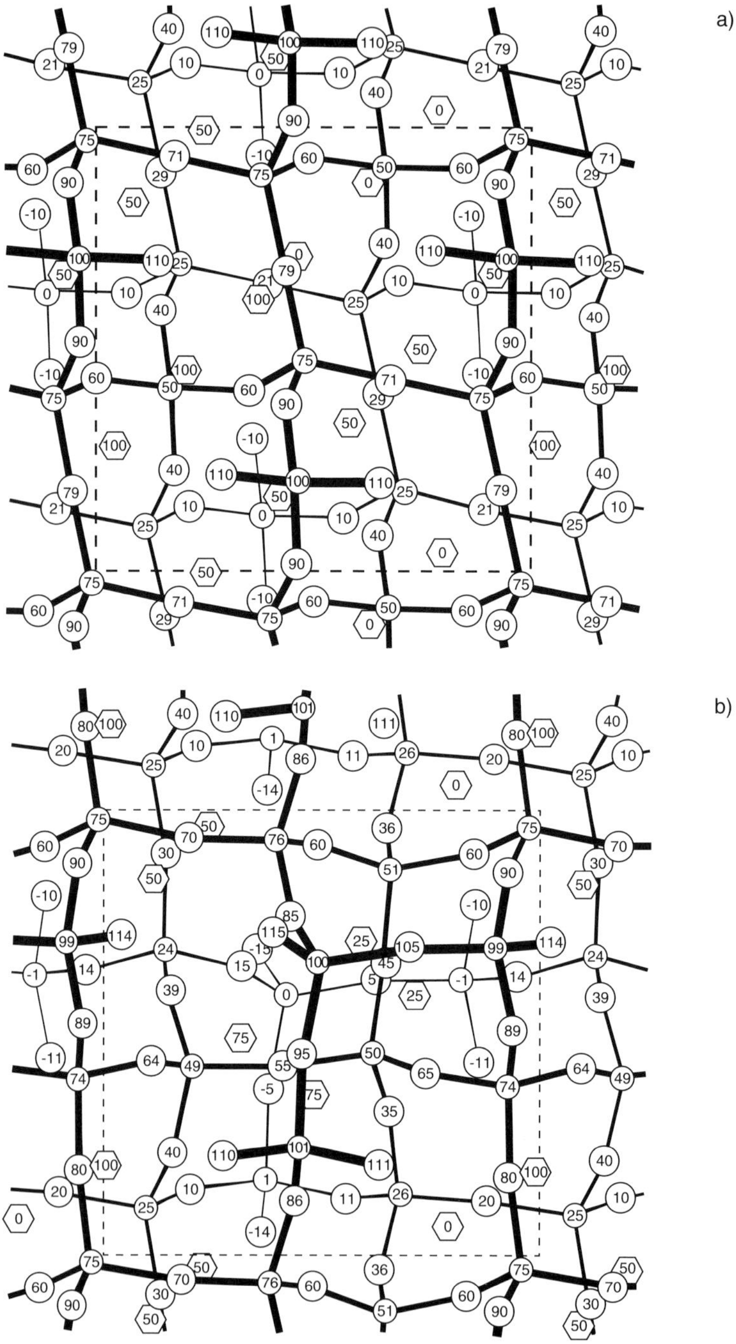

Fig. 2.3. Similarity of the h-quartz (**a**) and keatite (**b**) structures of $LiAlSi_2O_6$. Projection along the tetragonal and pseudotetragonal axis, respectively, from [2.10]. Small circles: (Al, Si). Large circles: O. Hexagons: Li. Numbers indicate relative elevation in the unit cell

known to have any range of thermodynamic stability at all. Yet they can, and often do, crystallize readily from supercooled liquids and can be so persistent metastably that they are useful in technical applications even at elevated temperatures.

The pseudo-binary system $LiAlO_2$-SiO_2, which is shown in Fig. 2.4 (modified after [2.13]), illustrates this quite well: the low-temperature phases eucryptite, spodumene, and petalite do not have framework structures, they cannot be crystallized directly from supercooled liquids of corresponding composition. The framework-type phases keatite s.s., high-quartz s.s., and β-eucryptite are the stable phases over a broad compositional range at high temperatures. They start melting at temperatures in the 1350–1450 °C range, incongruently in the case of β-eucryptite. At large undercooling, the first phase to crystallize from any melt in the composition range $LiAlSiO_4$ up to about $LiAlSi_4O_{10}$ is always a high-quartz s.s.. If their composition is in the keatite s.s. stability range, these phases generally require temperatures above 900 °C and extended periods of time to transform into the stable phase. It should be mentioned here that there is considerable inconsistency in nomenclature. In this paper we use the well-defined mineral names eucryptite, spodumene, and petalite for the low-temperature phases of $LiAlSiO_4$, $LiAlSi_2O_6$, and $LiAlSi_4O_{10}$. The term β-eucryptite, though not in line with mineralogical

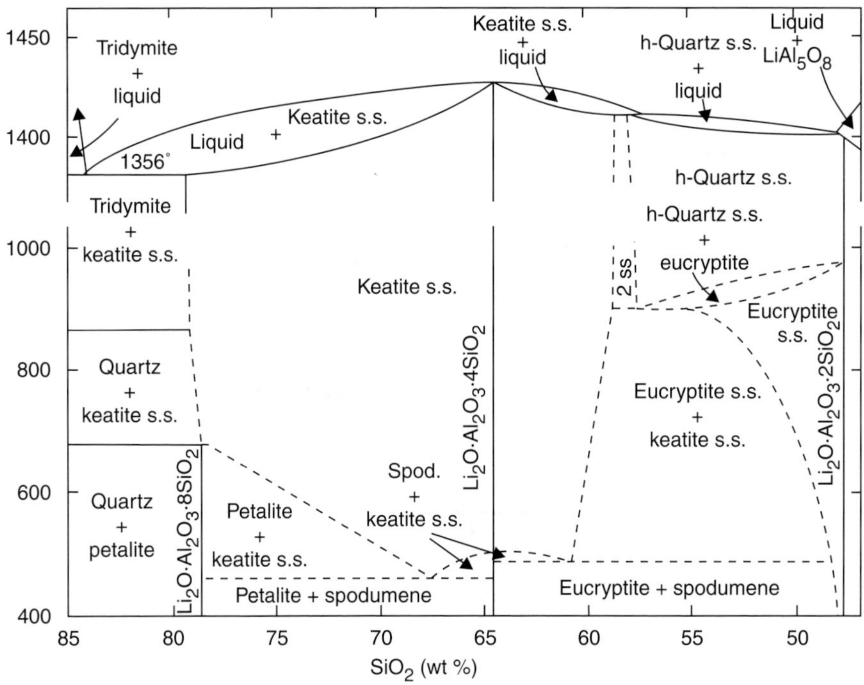

Fig. 2.4. System $Li_2O{\cdot}Al_2O_3{\cdot}2SiO_2$-$SiO_2$, after [2.13]

terminology, is in general use for the high-temperature form of $LiAlSiO_4$ with a quartz-type framework structure and, typically, ordered Al/Si distribution. With higher SiO_2 contents, the Al/Si order is lost. In order to emphasize the structural relationship, such phases are termed high-quartz s.s. here. In the literature, the terms β-quartz s.s. (correct), β-eucryptite s.s. (incorrect because of lack of Al/Si order), and silica-O (incorrect because this suggests some polymorph of SiO_2) are also used. For the stable, high-temperature form of $LiAlSi_2O_6$ the term β-spodumene is often used. As this phase and isostructural solid solutions bear no relationship with the pyroxene structure of spodumene, the term keatite s.s. is preferred here.

The thermodynamic stability of the high-quartz-type and, particularly, the keatite-type alumino-silicates is much reduced if part of the Li ions is replaced by the divalent ions Mg or Zn. Figure 2.5 shows this for the system SiO_2-$LiAlO_2$-$MgAl_2O_4$ [2.14]. Cordierite, $Mg_2Al_4Si_5O_{18}$, and the very stable spinel phases, $MgAl_2O_4$ and $ZnAl_2O_4$, are formed instead. Consequently, no fields of stability are found for high-quartz-type or keatite-type alumino-silicates in the systems $MgAl_2O_4$-SiO_2 or $ZnAl_2O_4$-SiO_2. As in the system $LiAlSiO_4$-SiO_2, however, metastable crystallization of high-quartz s.s. phases occurs readily from strongly supercooled liquids in the whole compositional range (Li, $Mg_{0.5}$, $Zn_{0.5}$)$AlSi_xO_{2x+2}$ with $1 \leq x \leq 5$, the upper limit being somewhat uncertain. The binary solid solutions $MgAl_2O_4$-SiO_2 have been studied by *Schreyer* and *Schairer* [2.15], those from the $ZnAl_2O_4$-SiO_2 system and more complex ones in detail by *Petzoldt* [2.16] and *Ray* and *Muchow* [2.17]. Many commercial low-expansion glass ceramics are based on them. (Such phases with high-quartz-type structure are sometimes called

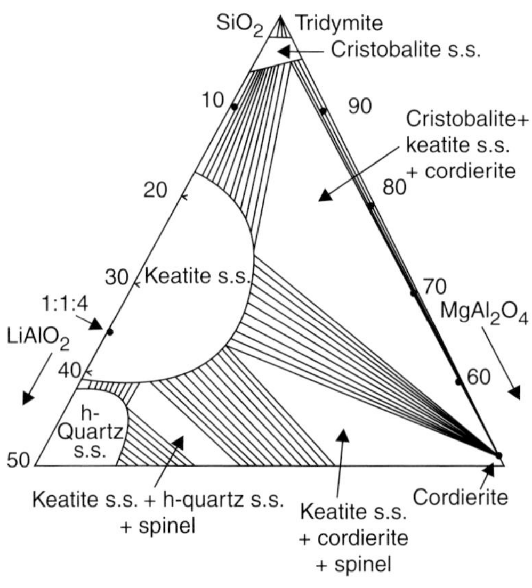

Fig. 2.5. Isothermal section from the system SiO_2-$LiAlO_2$-$MgAl_2O_4$ at 1230 °C, modified after [2.14]

"μ-cordierite" although, like in the case of β-spodumene, they are not related to the cordierite structure.)

A typical time–temperature transformation diagram for a glass ceramic containing, beside nucleating and fining agents, essentially the components SiO_2, $LiAlO_2$, $MgAl_2O_4$, and $ZnAl_2O_4$, is shown in Fig. 2.6. The considerable thermal persistence of the metastable high-quartz s.s. is apparent.

In contrast to most other alumino-silicates, the high-quartz s.s. phases, including the limiting case of β-eucryptite, allow replacement of a large part, i.e., up to 70%, of their SiO_2 content by $AlPO_4$ [2.19, 20]. In spite of the close similarity in the crystal chemistry of $AlPO_4$ and SiO_2, this substitution is generally very limited in framework silicates.

Further components, such as $Li_2[ZnO_2]$, $Li_2[BeO_2]$, or even $Al_{1/3}[AlO_2]$, have been reported as possible substitutions for $Li[AlO_2]$ in metastable h-quartz s.s., in concentrations between 10 and 25 mol% [2.14].

In the keatite-type alumino-silicates, possibilities for chemical substitution seem to be much more limited, even in their metastable state. In [2.21] it was found that 40 mol% of the Li_2O content in solid solutions of $LiAlSi_xO_{2x+2}$ could be replaced by MgO. However, the Al_2O_3 content in the same series could be partly (up to 25%) substituted by B_2O_3 and fully by Ga_2O_3. Partial substitution of Al by Ga and Si by Ge has also been found in β-eucryptite [2.22]. When the ranges of possible solid solutions of both structural series are compared, one has to bear in mind that the keatite s.s. can generally not be crystallized at large undercooling, because the metastable high-quartz series crystallizes first. It is quite conceivable that, under conditions of large undercooling, if they could be realized, the keatite structure were just as good solvent as the high-quartz structure.

In fact, the keatite structure appears to be much more flexible than the high-quartz structure. This has been shown by low-temperature ion exchange

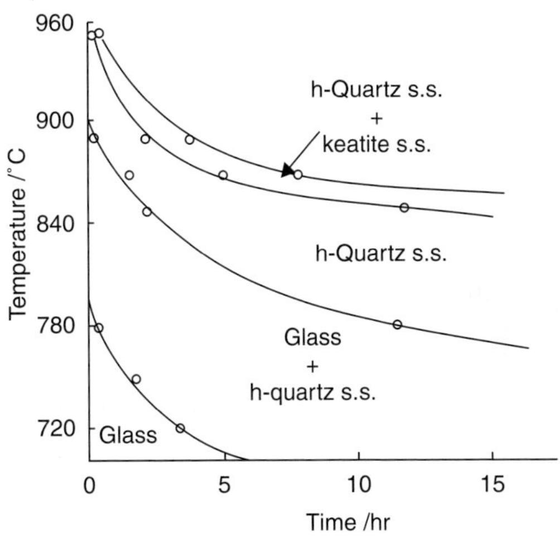

Fig. 2.6. Time–temperature transformation diagram for a glass ceramic from the $LiAlO_2$–$MgAl_2O_4$–$ZnAl_2O_4$–SiO_2 system after [2.17]

experiments. There, under conditions far from equilibrium and with strong chemical driving forces, the Li ions in keatite-type $LiAlSi_2O_6$ could be exchanged against H, Na, or even K ions without breakdown of the framework, whereas this exchange was much more limited in the corresponding high-quartz phase [2.23, 24].

2.1.3 Thermal Expansion

Large compositional areas of the solid solutions described above, particularly with the high-quartz structure, have very low or even negative thermal expansion. This makes such phases highly useful in low or zero thermal expansion glass ceramics. Table 2.1, compiled from various sources, gives lattice constants and coefficients of thermal expansion for selected representative compositions and both types of structures. In general it is observed that expansion coefficients vary only mildly, if at all, with temperature. Strong anisotropy of thermal expansion is characteristic for most members of the series. In some cases ($Zn_{0.5}AlSi_2O_6$ in Table 2.1), thermal expansion is negative in all directions.

Particularly the thermal expansion of β-eucryptite has been investigated by many authors. It appears that the expansion characteristics depend somewhat on the Al/Si order and possibly on other influences of thermal history, but the data given in Table 2.1 can be considered as typical. It should also be pointed out that coefficients of thermal expansion measured on polycrystalline aggregates by dilatometry can give different values due to internal stress and microcracking if the expansion of the crystals is highly anisotropic [2.27]. For β-eucryptite, several models have been proposed to explain the thermal expansion characteristics.

Table 2.1. Lattice constants and coefficients of thermal expansion of selected h-quartz-type and keatite-type alumino-silicates

Composition	Lattice constant Å		Thermal exp. coefficient 10^{-6}/K, 20–800 °C			Ref.
	a	c	α_a	α_c	α_V	
h-quartz-type						
$LiAlSiO_4$	10.497	11.200	8.6	−18.4	−0.4	[2.5]
$LiAlSi_2O_6$	5.212	5.457	1.0	−6.3	−2.1	[2.25]
$Zn_{0.5}AlSi_2O_6$	5.220	5.460	−0.5	−6.0	−2.1	[2.25]
$Mg_{0.5}AlSi_2O_6$	5.171	5.343	2.7	1.3	2.2	[2.25]
keatite-type						
$LiAlSi_2O_6$	7.545	9.156	−3.0	7.9	0.6	[2.26]
$NaAlSi_2O_6$	7.483	9.629	−0.9	3.4	0.5	[2.26]
$KAlSi_2O_6$	7.429	9.984	2.8	1.3	2.3	[2.26]

In 1959 *Gillery* and *Bush* [2.27] pointed out that the presence of interconnected helices of $(Si,Al)O_4$ tetrahedra in the β-eucryptite structure might be responsible for the expansion anisotropy: the helices run parallel to the hexagonal axis (c) in the high-quartz structure. As the helices are fixed within the structure, their thermal expansion was thought to cause torque stress in the sense of unwinding the helices. It was shown that this stress would contract the length of the helices and expand their diameter if the material building the helices had a ratio of Young's modulus E to bulk modulus K larger than 2. Large anisotropy of thermal expansion and c axis contraction could then be immediate consequences for β-eucryptite as well as other high-quartz-type phases.

The ideas of Gillery and Bush were taken up again in 1974 by *Moya* et al. [2.28]. They showed that the negative thermal expansion of β-eucryptite would require its isothermal linear compressibility (X) parallel to the c axis to be negative if the thermal expansion was dominated by elastic effects. The same group (*Hortal* et al. [2.29]) then measured compressibilities on β-eucryptite single crystals and found $X_{\|} = -(1.13 \pm 1) \times 10^{-13}\,\mathrm{cm^2/dyne}$ and $X_{\perp} = +(22.4 \pm 6) \times 10^{-13}\,\mathrm{cm^2/dyne}$.

So there is indeed a large anisotropy in the elastic properties, and the compressibility parallel to the c axis is probably negative, which led Hortal et al. to the conclusion that elastic effects govern the expansion behavior of β-eucryptite.

Quite a different approach was taken by *Schulz* in 1974 [2.30]. Crystal structure analyses (see Sect. 2.1.1) had shown that at room temperature the Li ions occupy sites with a tetrahedral coordination of oxygen atoms within the quartz channels. It had also been shown that the Li distribution about these tetrahedral sites changes from a rather ordered one at room temperature to a more disordered one at higher temperatures [2.5]. Redistribution of the Li ions occurs by rapid diffusion within the channels, which is also evidenced by the large ionic conductivity of β-eucryptite parallel to the channel direction [2.31]. In passing from one tetrahedral site to the next, a Li ion must move through a site with octahedral oxygen coordination. These sites are normally not occupied by Li ions at room temperature. They are, however, occupied by Mg ions in the Mg alumino-silicates with high-quartz-type structure [2.11]. The – unoccupied – octahedral sites in β-eucryptite have two oxygen atoms at the same z level with distances (1.84 Å) smaller than normal octahedral Li–O distances (2.00–2.41 Å). The remaining four oxygen atoms are much farther away (2.67 Å). Occupation of these sites by Li ions would be expected to push out the two close oxygens and attract the four others, with the consequence of a widening of the a axis and a contraction of the c axis. In Schulz's model the fractional occupancy of the octahedral sites is assumed to increase with temperature. The consequences for the lattice constants can then be calculated from structural data:

$$\Delta a = 4.46(N/M)\,\Delta d\ , \tag{2.1}$$

$$\Delta c = 11.47(N/M)\,\Delta d\ , \tag{2.2}$$

where N is the number of Li ions in octahedral coordination, out of a total number M of Li ions in the crystal, Δd is the increase of the distance between the two close oxygen atoms upon Li occupation of the site, and Δa, Δc are the resulting lattice-constant changes. With an estimated $\Delta d = 0.1$ Å, a change in the octahedral-site occupation of about 16% would give lattice-constant changes in agreement with the measured ones for a temperature increase from room temperature to 1000 °C.

Upon extended heat treatment, lattice constants and thermal expansion of β-eucryptite change slightly from those at room temperature. If these changes are assumed to be caused by increased disorder and a concomitant partial occupation of the octahedral sites even at room temperature, then the model by Schulz predicts the observed changes correctly. It is also in agreement with the fact that the Mg alumino-silicates have positive thermal expansion (because the Mg ions already occupy the octahedral sites at room temperature).

More recently, the crystal structure of β-eucryptite has been studied by single-crystal neutron diffraction, which gives the most reliable information on Li location and site occupancies. *Guth* made his analysis at room temperature and at 530 °C [2.6]; *Steinmann* worked at 767 °C [2.7]. Their results confirm that the differences in occupation of the Li sites with tetrahedral coordination become smaller with increasing temperature. Significant site occupation for the octahedral position could not be detected even at 767 °C. This casts some doubt on the validity of Schulz's model, but does not strictly disprove it because the occupancy of 10% required by the model at 767 °C may be too small to be detected even by neutron diffraction.

The neutron diffraction studies confirmed earlier X-ray results about other important structural changes with temperature [2.5]: the mean Si–O and Al–O tetrahedral distances were found to decrease continuously with temperature, whereas the Li–O distances increased; see Table 2.2. The decrease

Table 2.2. Average apparent bond distances (Å) in β-eucryptite and quartz as a function of temperature

	25 °C	530 °C	600 °C	767 °C
β-eucryptite				
Si–O	1.616	1.604		1.606
Al–O	1.733	1.727		1.722
Li–O	2.019	2.063		2.060
Ref.	[2.6]	[2.6]		[2.7]
Quartz				
Si–O	1.607		1.594	
Ref.	[2.32]		[2.32]	

in the Si–O and Al–O distances is comparable to the decrease in the Si–O distance measured in quartz, also listed in Table 2.2. In fact, zero expansion or slight apparent contraction of T–O bond lengths with temperature has been found in high-temperature structural analyses of many silicates. *Hazen* and *Finger* compiled data for silicates and other compounds and evaluated them statistically [2.33]. The large positive thermal expansion of the Li–O bonds in β-eucryptite is also in line with their findings for electrostatically weak M–O bonds.

In framework silicates, the apparent thermal contraction of the T–O bonds has been related to a transverse vibration of the oxygen atoms normal to the T–O–T bonding plane [2.34, 35]. Increased amplitudes of this mode, on time average, would increase the bond distance, which is compensated for by a movement of the Si atoms towards the oxygen. Consequently, if bond lengths are defined as distances between the centers of gravity of vibrating atoms, there is a shortening of bond distances with temperature.

The measured negative thermal expansion of high-quartz [2.36] and possibly also that of keatite (up to 300 °C) [2.37] are direct consequences of this bond-length shortening. In α-quartz the effect of bond-length shortening is masked by the tetrahedral rotation that causes a large positive expansion with temperature.

It can be supposed then that the thermal expansion of framework alumino-silicates, in the absence of phase transformations, is determined by the antagonistic effects of the contraction of the TO_4 tetrahedra and the expansion of the MO_x polyhedra of the cations M.

For the high-quartz-type and keatite-type phases of composition $MAlSi_2O_6$ listed in Table 2.1, the approach of polyhedral thermal expansions has recently been used for structural modeling [2.38]. Bond lengths based on ionic radii and bond expansion coefficients from the compilations in [2.33] were used to calculate idealized structures of the appropriate framework topology for room temperature and some high temperatures by least-squares refinement. Both the lattice constants at room temperature and the coefficients of thermal expansion that result from these calculations are in good qualitative agreement with measured data. Table 2.3 shows this for the axial coefficients of thermal expansion. As can be seen, the signs and the anisotropy of thermal expansion are correct in most cases. The positive thermal expansion of quartz-type $Mg_{0.5}AlSi_2O_6$ is greatly exaggerated in the model. The model structure, in this particular case, is α-quartz-like, in agreement with diffraction data [2.11], but with deviations from the high-quartz structure that are much more pronounced than those found experimentally. The distortions and the concomitant positive thermal expansion are direct consequences of the octahedral coordination of the Mg ions in the quartz channels, as has also been pointed out in [2.39].

In summary, the full range of thermal expansion characteristics encountered in α-quartz-type and keatite-type alumino-silicates appears to result

Table 2.3. Comparison of measured axial coefficients of thermal expansion and values calculated from computer-generated model structures for h-quartz-type and keatite-type alumino-silicates

	Thermal expansion coefficients (10^{-6}/K) measured (Table 2.1)		calculated [2.38]	
Phase	αa	αc	αa	αc
h-quartz-type				
$LiAlSi_2O_6$	1.0	−6.3	2.6	−5.6
$Zn_{0.5}AlSi_2O_6$	−0.5	−6.0	−0.1	−3.3
$Mg_{0.5}AlSi_2O_6$	2.7	1.3	13.1	22.6
Keatite-type				
$LiAlSi_2O_6$	−3.0	7.9	−1.2	7.0
$NaAlSi_2O_6$	−0.9	3.4	−1.2	5.8
$KAlSi_2O_6$	2.8	1.3	1.5	4.0

from the structural response to thermal bond-length changes of the T–O and M–O bonds. These changes, per se, do not differ appreciably from those found in many other structures.

2.1.4 Conclusions

Wide ranges of alumino-silicates exist with both the high-quartz and keatite structure, although areas of thermodynamic stability are rather limited and confined to elevated temperatures. The corresponding SiO_2 polymorphs can be considered as end-members. Large compositional areas in both series have very low or even negative thermal expansion, these areas also include the SiO_2 phases.

Specific models that have been suggested to explain the mechanism of thermal expansion of β-eucryptite, $LiAlSiO_4$, do not seem applicable to related phases or to the SiO_2 end-members.

High-temperature structural determinations on a large variety of crystals have revealed that coefficients of thermal expansion of a given type of cation-anion bond are fairly constant and depend inversely on the electrostatic bond strength. Tetrahedral (Al,Si)–O bonds in framework silicates, in particular, typically have a slightly negative or zero expansion coefficient.

If these bond expansion coefficients are used, computer modeling of the structures of various end members of the high-quartz-type and keatite-type alumino-silicates yields coefficients of thermal expansion in reasonable qualitative agreement with the measured ones.

Both the high-quartz and the keatite structure contain more than one possible site for the non-framework cations. Size and oxygen coordination preferences of the cations determine site occupations, which in turn exert significant influences on the framework geometry and thermal expansion.

This allows the thermal expansion to be tailored within wide ranges in solid solutions containing more than one cation.

2.2 Nucleation in Parent Glasses for Lithia Alumino-Silicate Glass Ceramics

Ulrich Schiffner

Glass ceramics are materials that are partly in a glassy and partly in a crystalline state caused by controlled crystallization of appropriate parent glasses. The crystal phase generally has a volume part between 50% and 90%, the sizes of the crystals are about 20 nm up to some µm. This means that the density of the crystals is between 10^{10} and 10^{15} crystals per mm^3. To obtain such high crystal densities in glass ceramics, on the one hand, certain catalysts – so-called nucleating agents – are usually added to the parent glass to support crystallization, on the other hand, an exact temperature profile has to be kept up during the crystallization process of the glass. For this reason, a knowledge as exact as possible of the crystallization behavior of the parent glasses is necessary for the technological control of glass ceramic production. The crystallization process is determined by two steps: by nucleation to cause the so-called nuclei and by crystal growth to develop the microstructure of the glass ceramic. Usually the nucleation is of dominant significance since it is of essentially stronger influence on both the processing of the glass and the properties of the glass ceramic than is the rate of crystal growth. Exact knowledge of the nucleation process in parent glasses for glass ceramics, however, is often not sufficient, because the experimental determination of the nucleation is very difficult. On the one hand, parent glasses have a very strong nucleation tendency and the nucleation and the crystal growth frequently show a strong overlap. On the other hand, the formation of nuclei is usually not accessible through experiments because of the small size of nuclei.

In the following, a survey is given about the phenomenon of nucleation including a short review of the historical development of investigations of nucleation. (Extensive presentations of the theory of nucleation have repeatedly been described in the literature [2.40, 41].) Following this, the determination of nucleation kinetics in a parent glass of a commercial lithia alumino-silicate glass ceramic will be presented (for references to this topic see [2.42–44]).

The first systematic investigations of the crystallization of glasses were performed by *Tammann* [2.45, 46]. He discovered that the transformation of a melt to a crystalline state consists of two different processes – the nucleation and the crystal growth. When a melt is cooled below the melting point, crystals do not form spontaneously in the entire volume. At first, submicroscopic crystalline aggregates of a definite size, i.e., the nuclei, must form, and only then can crystal growth occur upon these nuclei. The typical curves in

Fig. 2.7 show the dependence of nucleation rate and crystal growth on temperature. It is recognized that, on the one hand, high nucleation rates and, on the other hand, an overlap between both curves that is as small as possible are necessary to obtain as many nuclei as possible in the parent glass, on which crystals can later on grow at higher temperatures.

Satisfying theoretical statements on nucleation and crystal growth were found only some time after Tammann's experimental investigations. The basis of the classical nucleation theory is traced back to the ideas of *Volmer* and *Weber* [2.47] and *Becker* and *Döring* [2.48]. At first they characterized simple nucleation phenomena as the condensation of pure gas phases. Later on, *Turnbull* and *Fischer* [2.49] transferred the nucleation theory to the phase transformation liquid/solid and described the solidification of metal melts.

According to the classical nucleation theory, nucleation is perceived as a probability process. The formation of a crystalline phase in a metastable melt is characterized by the competition of two energy terms. The gain in energy due to the phase transformation is opposed by the energy costs for the creation of the interface between the melt and the new phase. Figure 2.8 shows the fundamental dependence of the Gibbs' free energy of nucleation

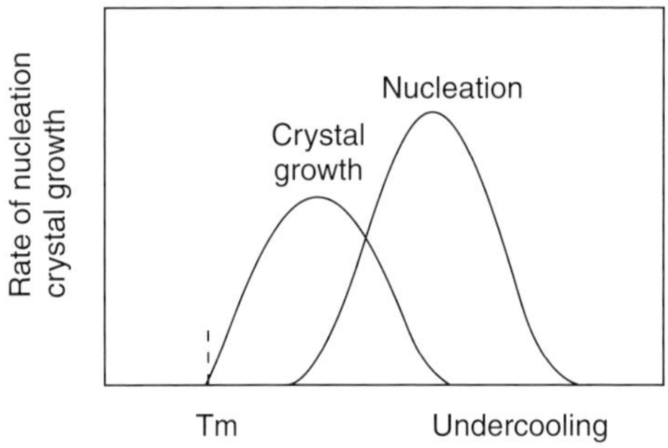

Fig. 2.7. Schematic dependence of the nucleation rate and crystal growth on the undercooling of a melt (T_m= melting point)

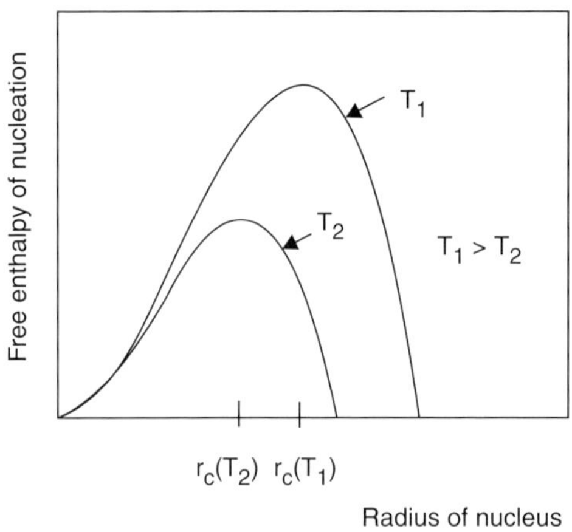

Fig. 2.8. Schematic dependence of the free enthalpy of nucleation on the radius of the nucleus at two fixed temperatures

on the radius of the nucleus at two fixed temperatures. One can see that the growth of crystalline aggregates is at first associated with an increase in Gibbs' free energy. With decreasing temperature, i.e., with increasing undercooling of the melt, the maximum of the Gibbs' free energy is reduced and nuclei form more easily. After a certain size is reached – called the critical radius – any further growth leads to a decrease in enthalpy. Aggregates with a radius smaller than the critical radius are usually called embryos, those with a radius greater than the critical radius stable nuclei. Embryos can form in spite of a cost in Gibbs' free energy in the beginning, because local molecular fluctuations occur continuously in an equilibrated system. Such fluctuations represent a steady alternating transport of molecules and enthalpy from one location in the system to another. Thus microregions form whose density and arrangement of particles no longer correspond to the homogeneous matrix, but may be considered as parts of a new phase. It is these microregions which are the starting points for the formation of embryos. The number of forming nuclei is determined by the number of embryos surpassing the critical size per time unit; and is called the nucleation rate. It strongly depends on temperature with a maximum, as presented by the Tammann curve in Fig. 2.7.

Up to this stage, the classical nucleation theory described the so-called steady-state nucleation as an equilibrium process independent of time. Then *Zeldovich* [2.50], *Frenkel* [2.51], and later on *Kashchiev* [2.52] extended the nucleation theory by studying the time dependence of the nucleation rate. After a change in temperature the nucleation rate corresponding to the new temperature does not occur immediately, but a certain time is required for establishing a new equilibrium rate. Nucleation during this time is called non-steady-state nucleation. The non-steady-state nucleation rate approaches the steady-state nucleation state with increasing time.

In a pure melt, embryos form at any site with equal probability. This nucleation is called homogeneous nucleation. However, in reality a melt contains many types of accidental impurities. These impurities can act as nuclei by reducing the energy which is necessary to create the interface between the melt and the new phase. This type of nucleation is called heterogeneous nucleation.

The mechanism of heterogeneous nucleation is the basis for the production of glass ceramics. In the 1950s *Stookey* [2.53] discovered that a glass can be converted into a uniformly fine-grained material by introducing submicroscopic catalyst crystals in a high degree of dispersion into a glass and by subjecting the glass to an appropriate annealing process. At first, photosensitive glasses were transferred into glass ceramics by this catalyzed crystallization. Small amounts of copper, silver, or gold were added to the glass as nucleating agents. They were precipitated in the form of very small crystals upon irradiation of the glass with ultraviolet light. In this way, Stookey found some important criteria for effective nucleating agents. These are good solubility in the glass at melting and forming temperatures, a high supersaturation on cooling, a low activation energy for diffusion, a low interfacial energy between the glass phase and nucleating agent, and, at last, similar crystal structures of the nucleating agent and the newly forming crystal phase.

The successful application of photosensitively crystallized glasses led to a large number of works to examine other glass systems and other nucleating agents for glass ceramic production. Some other nucleating agents as certain oxides, sulfides, and fluorides were detected to be effective catalysts. A detailed survey has been given by *McMillan* [2.54]. Among these nucleating agents TiO_2 and ZrO_2 have become most important. Both oxides show a similar behavior and have been used as effective nucleating agents in parent glasses for lithia alumino-silicate glass ceramics. This type of glass ceramic has been introduced in a large range of technical applications because of its very low expansion coefficient and its possible transparency.

Another result of nucleation research was the discovery of the synergistic effect of mixed nucleating agents. *Sack* and *Scheidler* [2.55] and *Stewart* [2.56] showed that combinations of both nucleating agents TiO_2 and ZrO_2 result in a better nucleation of the parent glasses than does the same concentration of just one of them. They lead to glass ceramics with finer grained microstructures and a higher transparency. Finally, by investigating combinations also of other oxide nucleating agents *Müller* [2.57] found that the nucleation is favored if only one single nucleating-agent crystal phase can be observed beside the main crystal phase of the glass ceramic after crystallization. This one nucleating-agent crystal phase was suggested to be the solid solution phase of both nucleating agents used. It was concluded that the formation of solid solutions of nucleating agents is the basis of an optimal nucleation in the parent glass. In spite of the successful realization of catalyzed crystallization for the production of glass ceramics, the detailed mechanism of nucleating agents in glasses is still being investigated. This particularly concerns the parent glasses of complex compositions where the glass and the forming crystalline matrix phase are composed differently. A large amount of research has shown that it is too simple to imagine that small nucleating agent crystals precipitate during heat treatment and act as heterogeneous nuclei in the glass. In most cases – as in repeatedly investigated TiO_2 nucleated parent glasses – it was evident that the first stage of the nucleation process is a glass-in-glass phase separation caused by the incompatibility of the nucleating agent with other oxides in the glass. In this context *Maurer* [2.58] and *Beall* [2.59] investigated magnesia alumino-silicate parent glasses and observed a TiO_2-rich liquid phase in the form of a dispersion of very small droplets. Next, small TiO_2-containing crystals formed which where assumed to act as nuclei for the matrix crystal phase. *Doherty* [2.60] and *Vogel* [2.61] found similar results in lithia alumino-silicate parent glasses. It is suggested that the composition in the droplets rich in TiO_2 approaches the composition of a nucleus phase rich in TiO_2 during phase separation. Finally, this nucleus phase attains a kind of pre-crystalline state. In a recent work, *Maier* [2.62] was the first to observe the formation of the nucleus phase in the droplets enriched by the nucleating agent oxides after phase separation. Then the formed nuclei were the starting points for the crystallization of the matrix phase.

On the other hand, further authors [2.63–65] found the nucleation process initiated by a glass-in-glass phase separation, but they could not identify a crystalline phase of the nucleating agents before the crystallization of the matrix phase. In these cases small crystals rich in TiO_2 were often observed at the grain boundaries of the matrix crystals only after the crystallization process. In this way, many investigations of the nucleating-agent mechanism show very inconsistent results. A possible explanation for this might be that the analysis methods used have not always been sufficient for an exact identification of very small crystalline aggregates in a glass.

As mentioned before, the experimental determination of nucleation in parent glasses for glass ceramics is very difficult. A suitable method for a quantitative evaluation of nucleation is the so-called two-step treatment or nucleus development procedure already repeatedly documented [2.42, 43, 66]. Samples which have been heat-treated in a defined manner within the nucleation region are subjected to a second heat treatment at higher temperatures above the nucleation region, permitting a rapid growth of crystals on the nuclei. Subsequently, the number of crystals in the sample is determined and is assumed to correspond to the number of nuclei formed. Since the parent glasses for glass ceramics have extremely high nucleation rates compared with conventional glasses, nucleation studies require very short treatment times and exact temperature control. For this reason, Schott AG in Mainz has developed a special apparatus permitting two-step heat treatments for a few seconds with high precision [2.67].

In the following, some results of the investigations of nucleation kinetics in a parent glass for a commercial lithia alumino-silicate glass ceramic are presented. The composition of this glass is listed in Table 2.4. The samples used were cut from glass rods quenched in air. The number density of crystals obtained by a two-step treatment is plotted as a function of nucleation time for every fixed nucleation temperature. The conditions of the second heat treatment – the so-called treatment of development – were kept constant with $T_d = 1000\,°C$ and $t_d = 4\,min$ to obtain comparable results.

The curves – so-called nucleation isotherms – are shown in Fig. 2.9. The slopes correspond to the nucleation rates. The shapes of nucleation isotherms vary considerably for different nucleation temperatures. However, they are all characterized by the occurrence of three typical sections. The first section

Table 2.4. Composition of the investigated parent glass (wt%)

SiO_2	Al_2O_3	Li_2O	Na_2O	K_2O	MgO	ZnO	P_2O_5	TiO_2	ZrO_2	As_2O_3
55.4	25.4	3.7	0.2	0.6	1.0	1.6	7.2	2.3	1.8	0.6

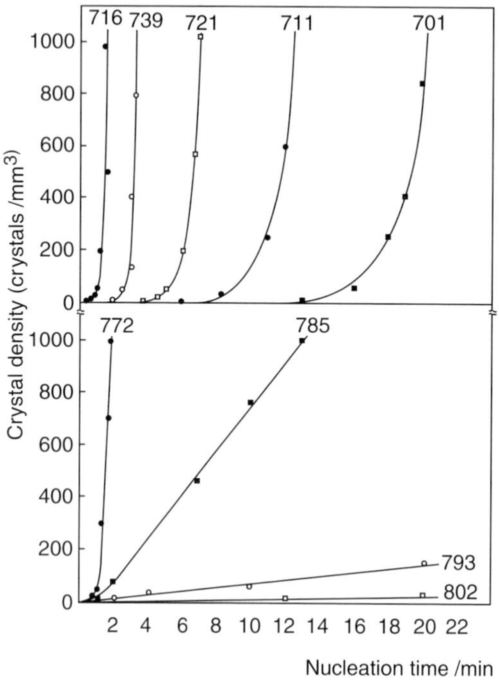

Fig. 2.9. Dependence of the crystal density on the nucleation time at various nucleation temperatures (in °C)

represents a delay without any effective nucleation taking place (nucleation rate is zero). In the second section the density of crystals increases almost exponentially (nucleation rate increases). Finally, in the third section the density of crystals increases linearly with time (nucleation rate is constant). The linear section can be recognized unequivocally only at high nucleation temperatures. The isotherms at lower temperatures could be followed only for a short period of time since the highest crystal density capable of being determined was soon exceeded because of the high nucleation rate observed.

The dependence of the temperature range of effective nucleation on time is presented in Fig. 2.10, where crystal density is plotted as a function of nucleation temperature at constant nucleation times, based on the data of Fig. 2.9. The crystal density strongly increases with the nucleation time so that complete curves can no longer be obtained. At the same time the temperature range of nucleation widens as the lower limit of nucleation is shifted to lower temperatures. Conversely, the upper limit shows only little dependence on nucleation time.

So far, the results have demonstrated that nucleation is characterized decisively by non-steady-state effects. The typical shape of the isotherms illustrated in Fig. 2.9 is the result of the continuous approach of the non-steady-state nucleation rate to the equilibrium nucleation rate at any fixed temperature. The different shape of isotherms shows that the high-temperature and

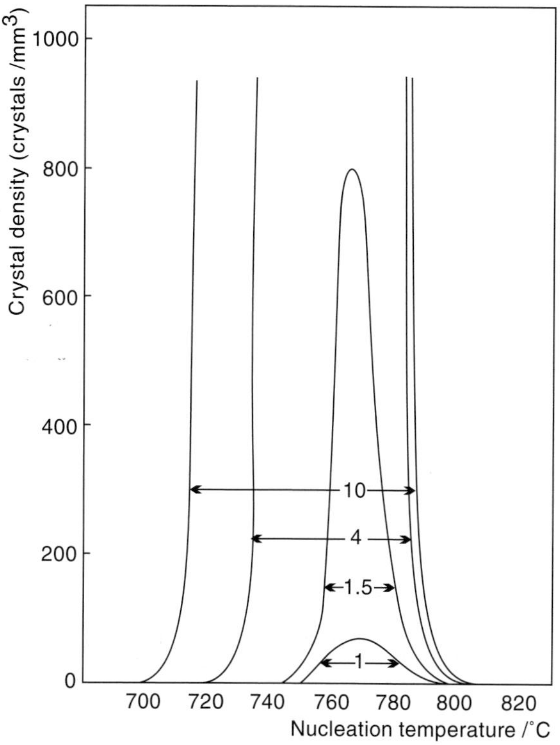

Fig. 2.10. Dependence of the crystal density on the nucleation temperature at constant nucleation times (in min)

low-temperature limits of the area of nucleation are caused by two different mechanisms. At low temperatures non-steady-state processes dominate. The delay times are very large and nucleation cannot be observed when the delay time is longer than the nucleation time in the experiment. At high temperatures, however, since the equilibrium state is attained rapidly, the limit of nucleation is determined by the decrease of the equilibrium nucleation rates to values below experimental observation.

The dependence of the nucleation on time and temperature can be presented in a lucid form by the time–temperature transformation (TTT) diagram. The amount transformed corresponds to the crystal density. All time–temperature conditions leading to the same crystal density are connected by lines. Figure 2.11 shows the TTT diagram derived from the data of Fig. 2.9. The density line $\rho = 1$ is the outermost boundary of the nucleation area. It indicates the conditions for the creation of the first effective nucleus. The characteristic shape of the density curves results from the interplay of the two different mechanisms mentioned above, which are responsible for the limitations of the nucleation area at high and low temperatures, respectively. This shape delineates a temperature range of minimum times for the formation of the first nucleus.

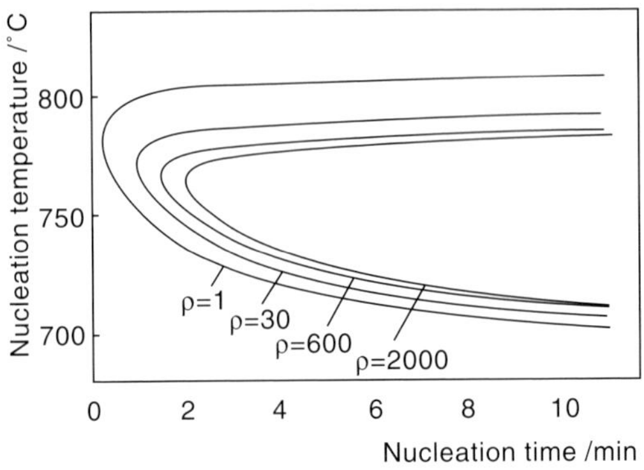

Fig. 2.11. Time–temperature transformation (TTT) diagram of nucleation showing several crystal density lines (ρ)

Knowledge of the TTT diagram and, particularly, of the $\rho = 1$ density line is of great importance for the forming and hot working processes of glasses with a strong tendency towards nucleation. From this diagram, schedules can be derived which avoid nucleation. For the forming process of the parent glass the main interest is in the upper limit of nucleation. Especially during casting the temperature should not fall below this limit. Otherwise, the creation of nuclei is inevitable and crystals will grow on these nuclei upon reheating the glass. The knowledge of the $\rho = 1$ density line in the temperature range where the delay times are very short is of particular interest for hot working processes. Reheating the glass without nucleation requires passing this critical temperature range within the shortest delay time.

The method of short-time heat treatment used so far allows the determination of the initial phase of nucleation. To determine the conditions for optimal treatments of the parent glasses in order to yield glass ceramics with high quality properties, usually with crystal densities of 10^{10}–10^{15} crystals per cubic millimeter, other methods have to be chosen. Preliminary experiments suggest that the light transmission of glass ceramics obtained by different nucleation conditions may serve as such a method in the particular system considered here. In order to achieve glass ceramics of high transparency, either very small crystals or close similarity of the refractive indices of crystalline and vitreous phase are required (Rayleigh's law of light scattering). Control of the indices is generally difficult in the case of glass ceramics. Therefore, adjustment of transparency can only be utilized as a relative measure of the crystals' sizes. Since the size is essentially determined by the amount of nuclei formed during the nucleation treatment, transmission data can supply an indication of the intensity of nucleation.

For the determination of transmission, glass rods were exposed to various nucleation treatment times and, subsequently, to an identical development treatment. For each sample cut from ceramized rods a transmission spectrum was taken on a double beam spectrometer. In Fig. 2.12 transmission at the wavelength of 450 nm is plotted versus the nucleation temperature for various nucleation times. The transmission value at 450 nm is used for comparison, since it presents the largest differences. Samples treated for less than 45 min are completely opaque, independently of the nucleation temperature. For some nucleation temperatures, transmission starts to be measurable with nucleation times exceeding 45 min and increases with rising nucleation time to a maximum which is not exceeded at longer times. At the same time the range of transparency broadens with the lower limit moving to lower temperatures and the upper limit remaining almost unchanged.

The dependence of the transmission on nucleation temperature and time shows great similarity with that of crystal density as presented in Fig. 2.10, demonstrating the close relation between the crystal density and transparency for this glass ceramic. To obtain transparency a certain crystal size and thus a certain minimum density of nuclei is needed. The nucleation time required for obtaining this density of nuclei is determined, on the one hand, by the duration of the non-steady-state effects of nucleation and, on the other hand, by the value of the equilibrium nucleation rate at the respective nucleation temperature. As delay times are short at temperatures near the upper limit of the range of transparency, the transparency may be observed here after short treatment times. In this case, however, an increasing nucleation time does not lead to higher degrees of transmission. This means that at these temperatures the nucleation rates are too low to yield crystal densities and corresponding crystal sizes leading to higher transmission. At temperatures below 700 °C transmission attains an identical maximum independent of the nucleation temperature if, corresponding to non-steady-state nucleation effects, the nucleation times are sufficient.

Figure 2.13 shows the dependence of the transmission on the nucleation temperature and time in a TTT diagram. The temperature and time conditions leading to equal transmission are again connected through lines. To illustrate the region of transparency in the entire area of nucleation, the limits of nucleation have also been inserted. For short nucleation times (up to about 10 min) results are used which are shown in Fig. 2.11. Boundaries for

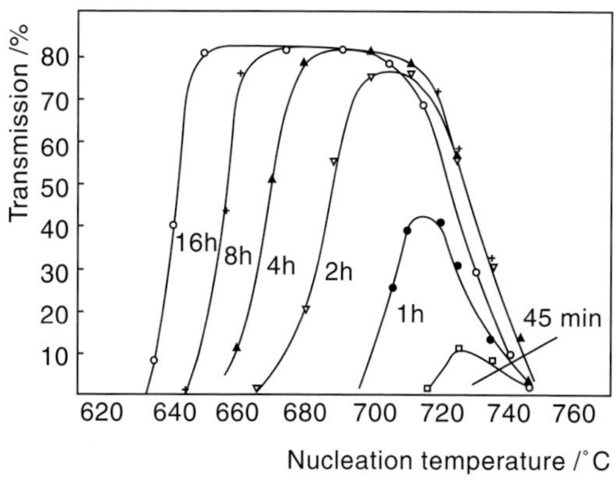

Fig. 2.12. Dependence of the transmission at $\lambda = 450$ nm on the nucleation temperature for various nucleation times

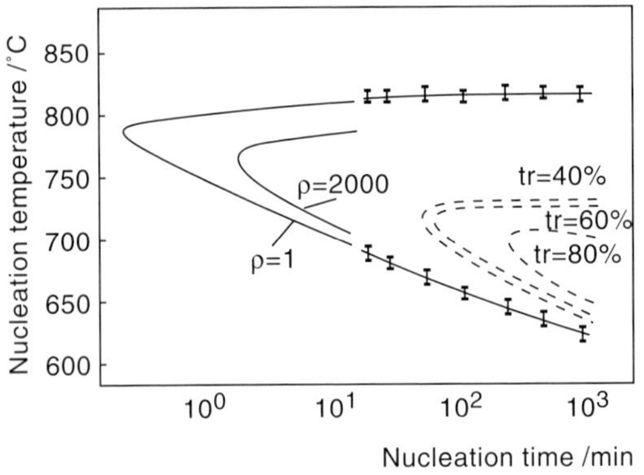

Fig. 2.13. TTT diagram of nucleation with crystal density lines and lines for various degrees of transmission (tr)

longer nucleation times were determined by the transmission experiments with gradient rods. The limit of nucleation was taken from the transition of the vitreous to the opaque crystalline zone.

Thus a TTT diagram of nucleation for a large time and temperature range is available for the investigated glass. From this diagram decisive conditions of nucleation can be derived for further treatments. With subsequent hot working these conditions lead to the creation of the first nucleus – particularly in the temperature range of the shortest delay times. Conversely, for ceramizing, these conditions assure optimum transparency.

So far the concentration of the nucleating agents TiO_2 and ZrO_2 has been constant in the investigated glass. In the following, the amounts of the two nucleating agents are varied in a glass of a similar base composition (Table 2.5) in order to study the influence of the nucleating agent concentration on the nucleation behavior of the glass. This was carried out in three experimental series. At first the TiO_2 content was varied between 0.5 and 3.0 wt%. In the second series the ZrO_2 content was varied between 0 and 2.0 wt% and, finally, the ratio of TiO_2/ZrO_2 was varied at a constant sum of weight. The compositions of all the glasses are listed in Table 2.6.

Figure 2.14 shows the isotherms of the TiO_2-varied glasses at a nucleation temperature of 740 °C for each one. It can be seen that, on the one hand, delay times increase, on the other hand, the rise of the isotherms decreases with decreasing TiO_2 content. This suggests that the reduction in the TiO_2 content slows down nucleation.

The TTT diagrams shown in Fig. 2.15 are obtained from the isotherms of each glass. For clarity only the $\rho = 1$ density lines are given. In contrast

Table 2.5. Composition of the base glass (wt%)

SiO_2	Al_2O_3	Li_2O	Na_2O	K_2O	ZnO	BaO	Sb_2O_3
67.6	22.5	3.8	0.6	0.3	1.5	2.5	1.1

Table 2.6. Compositions of glasses with varying nucleating agent content (wt%)

Variations in TiO_2						
TiO_2	0.5	1.0	1.5	2.0	2.5	3.0
ZrO_2	1.5	1.5	1.5	1.5	1.5	1.5
Base glass	98.0	97.5	97.0	96.5	96.0	95.5
Variations in ZrO_2						
TiO_2	2.5	2.5	2.5	2.5	2.5	2.5
ZrO_2	0.5	0.75	1.0	1.25	1.5	2.0
base glass	97.0	96.75	96.5	96.25	96.0	95.5
Variations in TiO_2/ZrO_2						
TiO_2	4.0	3.5	3.0	2.5	2.0	1.5
ZrO_2	0.0	0.5	1.0	1.5	2.0	2.5
Base glass	96.0	96.0	96.0	96.0	96.0	96.0

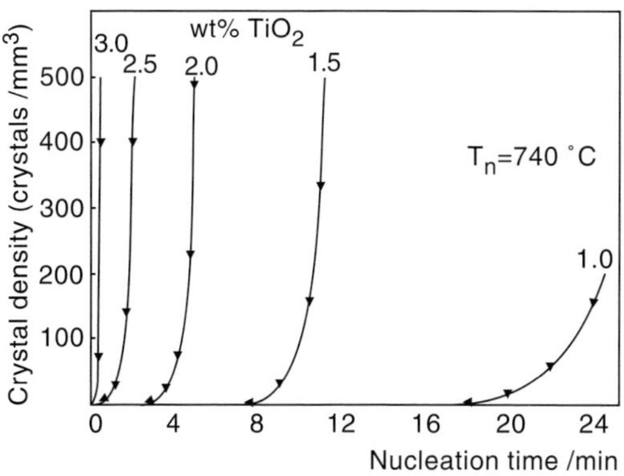

Fig. 2.14. Nucleation isotherms at $T_n = 740\,°C$ for glasses with varying TiO_2 content

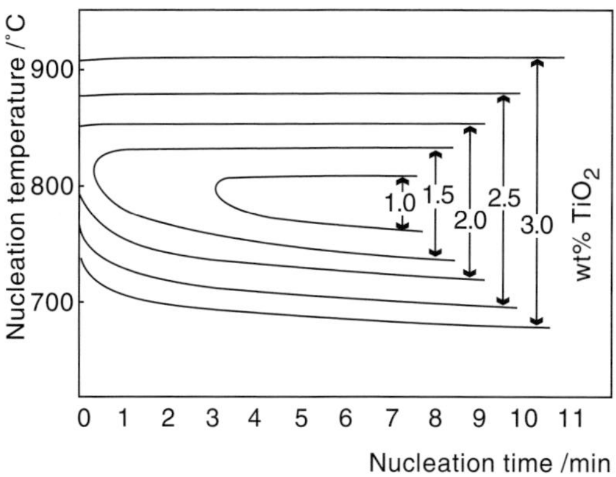

Fig. 2.15. TTT diagrams ($\rho = 1$ density lines) of glasses with varying TiO_2 content

to the glass investigated above, the delay times of glasses containing more than 1.5% TiO_2 are so short that they fall below the limit of observation for certain temperature intervals. This means that in these cases nuclei are formed as early as during the short heating process of some seconds. Furthermore, Fig. 2.15 shows that a significant diminution of the nucleation area occurs with decreasing concentration of TiO_2. On the one hand, delay times decrease; on the other hand, the upper limit of nucleation is shifted towards lower temperatures. For the glass containing 0.5% TiO_2 no nucleation is observed within the entire temperature range for a nucleation time of 20 min.

The nucleation isotherms of the glass with varying ZrO_2 content are seen in Fig. 2.16 at the same nucleation temperature as before. As in the case of the TiO_2 variations, the decrease in ZrO_2 content clearly slows down nucleation. Figure 2.17 shows the influence of the ZrO_2 concentration on the TTT diagram. Again, the reduction in concentration causes a narrowing of the boundaries of nucleation, and thus a decrease in the area of nucleation. No boundaries of nucleation could be determined for the two glasses containing 0.5 and 2.0 wt% ZrO_2. The glass with 0.5 wt% ZrO_2 shows no effective

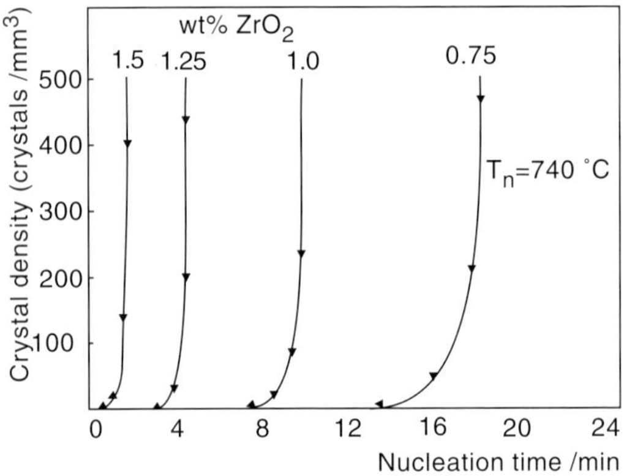

Fig. 2.16. Nucleation isotherms at $T_n = 740\,°C$ for glasses with varying ZrO_2 content

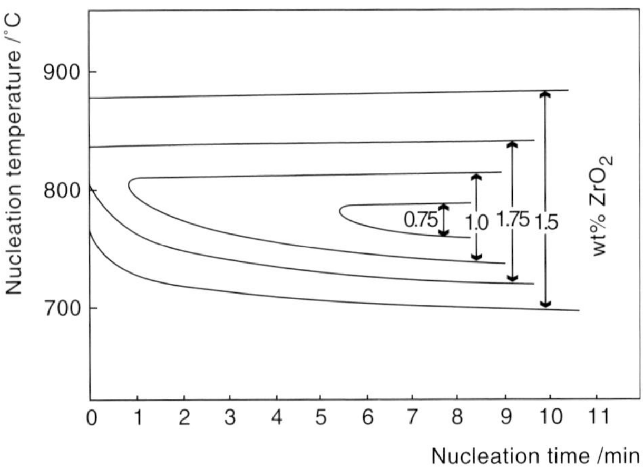

Fig. 2.17. TTT diagrams ($\rho = 1$ density lines) of glasses with varying ZrO_2 content

nucleation within the observation time. The glass containing 2.0 wt% ZrO_2, however, is so strongly nucleated in spite of quenching that crystal densities are far above the experimental limit even before nucleation treatment.

Finally, the isotherms of the glasses with a modified ratio of TiO_2 and ZrO_2 are shown in Fig. 2.18 for the nucleation temperature of 700 °C. The position and the shape of the isotherms show that starting from a ZrO_2-free glass an initial small replacement of TiO_2 by ZrO_2 slows down nucleation whilst further replacements accelerate the nucleation process. In Fig. 2.19 the TTT diagram is presented. There is no easily detectable dependence on concentration changes such as observed in glasses in which either TiO_2 or ZrO_2 alone is varied. Starting from the ZrO_2-free glass, replacement of 0.5 wt% TiO_2 by ZrO_2 causes the upper and lower nucleation boundaries to move closer to each other. An exchange of an additional 0.5 wt% shifts both

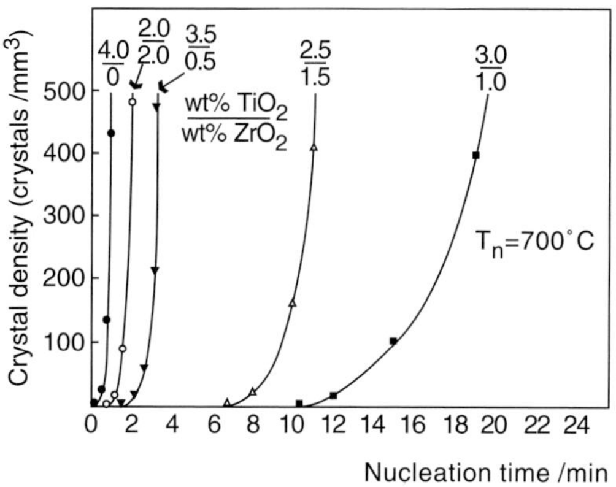

Fig. 2.18. Nucleation isotherms at $T_n = 700\,°C$ for glasses with a varying TiO_2/ZrO_2 ratio

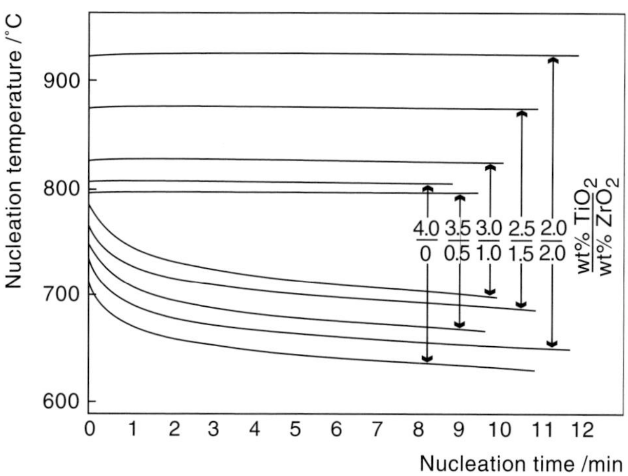

Fig. 2.19. TTT diagrams ($\rho = 1$ density lines) of glasses with a varying TiO_2/ZrO_2 ratio

boundaries towards higher temperatures, while the width of the nucleation area remains constant. Further replacement of TiO_2 by ZrO_2 finally leads to a continuous widening of the area, similarly to the effect of increasing just one nucleating agent. The glass containing 1.5 wt% TiO_2 and 2.5 wt% ZrO_2 is so strongly nucleated in the quenched condition that no boundaries of nucleation can be determined. In contrast to the behavior of glass with variations of either TiO_2 or ZrO_2, simultaneous changes in both components do not lead to a detectable shift of the nucleation area on the time axis. Thus, the delay times of all glasses are in certain – although different – temperature intervals below the limit of experimental detection.

The results of investigations with glasses with varying concentrations of a single nucleating agent show that a decreasing concentration causes both a slowing down of the nucleation process and a lowering of the upper limit of nucleation. It follows – as shown by the TTT diagrams – that these changes narrow the critical temperature interval in which delay times are short as well as significantly increase the shortest delay time. Since for forming and subsequent hot working a low upper nucleation temperature as well as a very long delay time are desirable, these results demonstrate that a reduction in the concentration of one nucleating agent improves the workability of the glass. In contrast, the exchange of the two nucleating agents in equal amounts just shifts the upper boundary of nucleation while the minimal delay time remains extremely short, independent of the ratio of the two nucleation oxides. Thus, these glasses are not well suited for subsequent hot working processes.

A comparison of the glasses in which either TiO_2 or ZrO_2 is varied shows that a change in ZrO_2 has the same but much stronger effect than the corresponding change in TiO_2. In the glasses with simultaneous variation of both nucleating agents, however, the stronger effect of ZrO_2 exhibits itself only above a certain minimum concentration of ZrO_2. Actually, the initial replacement causes a decrease in nucleation. This indicates that the nucleating agent must attain some minimum concentration to be effective. Thus, the decrease in nucleation brought about by replacing 0.5 wt% TiO_2 by ZrO_2 might be solely due to the decrease in TiO_2. A further exchange finally widens the nucleation area because of the stronger effect of ZrO_2 on nucleation.

The fact that ZrO_2 is more effective than TiO_2 is demonstrated more clearly if one considers that by exchanging weight percentages a larger molar amount of TiO_2 is replaced by a smaller one of ZrO_2. This means that in spite of a reduction of the sum of molar concentrations, nucleation is enhanced by ZrO_2. From this behavior it can be concluded that the smaller variation in ZrO_2 causes a larger change in the enthalpy of nucleation than does the larger variation in TiO_2. Besides explanations based on changes in the structure of the embryo phase due to shifts in the TiO_2-ZrO_2 phase diagram, an explanation of the stronger effect of ZrO_2 in comparison with TiO_2 might be provided by the different field strengths of the cations Ti^{4+} and Zr^{4+}. According to *Dietzel* [2.68], the Ti^{4+} cation is usually precipitated as the free oxide from a silicious melt because of its relatively high field strength (1.25), whereas the Zr^{4+} cation (0.78) can be precipitated as zirconium silicate ($ZrO_2 \cdot SiO_2$). It might be conceived that this silicate could be a more effective nucleating agent for a silicate phase than pure TiO_2.

2.3 Glass Ceramics Based on Lithium-Alumino-Silicate Solid Solution Crystals

Wolfgang Pannhorst

In Sect. 2.1 the main aspects of h-quartz and keatite solid solution formation of compositions of type $MAlSi_xO_{2x+2}$ have been discussed. The discussion was mainly concentrated on thermodynamically stable compositions although it was also described that metastable h-quartz s.s. phases can be obtained by devitrifying glasses, thereby extending the composition field for h-quartz s.s. crystal formation. In addition, Fig. 2.6 displays a TTT-diagram of the field of metastability for a glass ceramic with a composition in the $LiAlO_2$-$MgAl_2O_4$-$ZnAl_2O_4$-SiO_2 system according to [2.69].

Although the stable phases can be obtained by solid state reactions, i.e., ceramic technology, as well as by devitrification, it was the properties of the metastable phases which mainly attracted the scientists. The broader range of allowed composition variations was better suited to meet all the requirements of the envisaged products, even though the property of metastability itself poses restrictions on the choice of the final composition in some applications. Thus, the glass ceramic approach appeared to be superior to the ceramic approach and it is this approach which is described in the following. At the same time further discussion is restricted to h-quartz s.s. containing glass ceramics.

The development of a glass ceramic comprises two steps which are mutually dependent on each other: (1) the development of a base glass composition, and (2) a ceramization process by which the base glass is transformed into a glass ceramic with specified material properties. In principle there are two options for the development of low-expansion glass ceramics. One option is based on a crystalline phase which itself has a coefficient of thermal expansion (CTE) close to zero in the required temperature interval; in this case a corresponding composition has to be found and a ceramization process has to be developed so that the base glass transforms to almost 100% into the designed crystalline phase. The second option is based on a crystalline phase which itself has a negative CTE in the required temperature interval; in this case the glass ceramic has to be designed in such a way that it consists partly of that crystalline phase with negative CTE and partly of a glass phase (called residual glass phase) with positive CTE; the CTEs and the fractional volumes of both phases are adjusted to each other so that the CTE of the glass ceramic is close to zero [2.70].

In the development of low-expansion LAS (LAS = Li_2O-Al_2O_3-SiO_2) glass ceramics the second approach has always been favored. The requirements for melting a base glass of high quality as well as those for fine tuning the CTE during ceramization can be met more easily by this second approach.

Section 2.2 describes in detail how homogeneous, inclusion-free glasses of the extended LAS system can be transformed into homogeneous, inclusion-free glass ceramics by the ceramization process when proper amounts of nucleating agents have been added to the composition.

Section 2.2 also describes how the interplay between the amount of nucleating agents on the one hand, and the ceramization conditions on the other hand, determines the final properties of the glass ceramic. While this detailed account exemplified how optimized ceramization conditions can be derived once a base glass composition has been chosen, in this section we intend to provide some guidelines for the development of a new glass ceramic.

The quite broad composition field for metastable h-quartz s.s. formation as well as the possibility to design a low-expansion glass ceramic by mixing a crystalline phase with negative CTE in proper amounts with a glassy phase with positive CTE lead to an incalculable number of potential base glass compositions. Therefore, there is no straightforward process for finding the correct glass composition and its ceramization conditions to meet specific requirements but there exist always several solutions simultaneously; it is up to the intuition of the developer which solution is finally proposed.

The development of a glass ceramic composition always starts with a base glass composition taken from the area indicated in Fig. 2.20. The upper amounts of SiO_2 and Al_2O_3 or the sum of both components are often determined by the refractoriness of these components. To obtain compositions which are meltable with good glass quality in normal glass tanks, the viscosity at melting temperature should not exceed a few 100 dPa s. A further point to observe was pointed out by *Petzoldt* [2.71]. The glass ceramics are based on h-quartz s.s. crystals which do not undergo any high–low phase transition such as that known for pure quartz. If the SiO_2 content is chosen to be high, this phase transition may show up; this was demonstrated for the systems Li_20-Al_2O_3-SiO_2 and ZnO-Al_2O_3-SiO_2. The upper amount of Li_2O is in most cases determined by the CTE of the glass ceramic. To achieve a CTE close to zero for the glass ceramic that of the crystalline phase should not become too negative; remember that the volume CTE of $LiAlSi_2O_6$ is as low as $2.1\times10^{-6}\,K^{-1}$ (Sect. 2.1).

Starting from a three-component composition further modifications are possible by (a) substituting $2SiO_2$ in the crystalline phase by $ZnAl_2O_4$, $MgAl_2O_4$, or $AlPO_4$ or by (b) modifying the composition of the residual glass, Al_2O_3-SiO_2, by adding glass modifiers such as Na_2O, K_2O, CaO, or BaO. These are the components most often used; other components are considered to be too expansive. If the determination of a base glass composition is already a ten-parameter problem, its choice is even more complicated as it

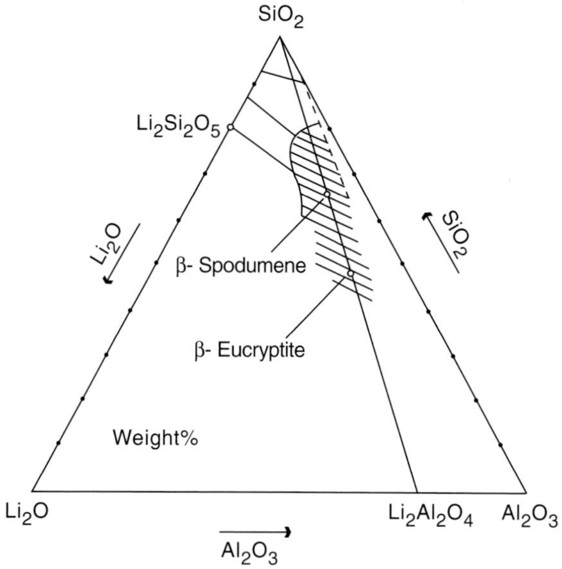

Fig. 2.20. Basic phase diagram for LAS-glass ceramics after *Smoke* [2.70], modified; hatched area indicates crystalline phases with low or negative volume expansion

also depends strongly on the ceramization conditions. Therefore, there exist no data of a systematic investigation of all possibilities. It is assumed that various laboratories have accumulated sufficient data in certain composition areas to optimize compositions within these areas by regression analysis. The way in which such optimization processes are principally performed is described below.

While one aspect of the development of a glass ceramic is the determination of specified properties, another aspect concerns the kinetics of the transformation of the base glass into the h-quartz s.s. containing glass ceramic and the stability of this phase. A first overview of the kinetics of this phase transformation is obtained by either differential thermal analysis (DTA) or differential scanning calorimetry (DSC). The position of the transformation peak indicates the temperature range in which volume crystallization proceeds rapidly. Figure 2.21 shows the DSC curve of the base glass of the glass ceramic Robax® taken at a heating rate of 5 K/min. The strong exothermic peak at about 838 °C is characteristic of the crystallization of the h-quartz s.s. crystals. In this temperature range the heating rate usually has to be reduced during ceramization to secure a uniform transformation within the whole glass ceramic object under development. In the case of very large objects for which a high temperature homogeneity during ceramization is important even the heat released by the exothermic transformation has to be taken into account. Figure 2.21 displays two more small exothermic peaks [2.72] which are discussed further below.

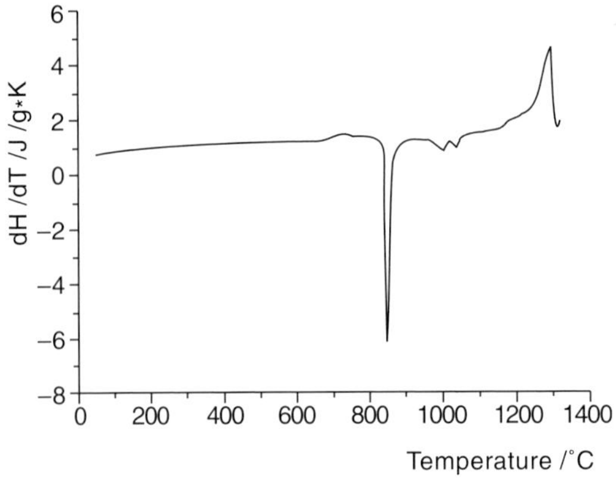

Fig. 2.21. DSC trace for Robax® base glass taken with a heating rate of 5 K/min

The peak temperature of the transformation depends highly on the choice of the nucleating agents [2.73]; it is shifted to lower temperatures when a combination of ZrO_2 and TiO_2 is chosen instead of either TiO_2 or ZrO_2 alone. In the case investigated in [2.73] the lowest temperatures are observed when a molar ZrO_2/TiO_2 ratio somewhat higher than 0.5 is chosen.

A low transformation temperature is important in the production of thin components such as kitchen articles. In this case small temperature inhomogeneities during ceramization are less important than the ability of the articles to maintain their shape. The heating rate during ceramization is, therefore, determined by the viscosity changes during ceramization; after increasing the temperature above the glass transition temperature it is desirable that the glass articles soon transform into glass ceramics with increased viscosity. An article which transforms at lower temperatures can thus be ceramized faster than one transforming at higher temperatures. The viscosity variation of a base glass composition during ceramization measured with a beam viscosimeter is shown in Fig. 2.22 for a heating rate of 5 K/min. During heating to 800 °C the glass behaves almost like a normal glass; there is only a small deviation from a normal glass viscosity curve just before the steep increase in viscosity starts. When the crystallization proceeds the viscosity increases rapidly by two to three orders of magnitude, thereby ensuring sufficient stiffness of the body during further heating.

Another aspect to be considered during ceramization is the density change occurring during the phase transformation. During the transformation the density increases roughly by 3 vol% resulting in a linear shrinkage of about 1%. Figure 2.23 shows the length changes of a specimen during ceramization when heated up with 5 K/min in a dilatometer. Although the curve has not been corrected for the length changes due to either thermal expansion or slight deformations of the specimen at lower viscosities, the figure clearly displays the large length change in the temperature interval 730–850 °C in which the main transformation occurs. As the shrinkage is accompanied by

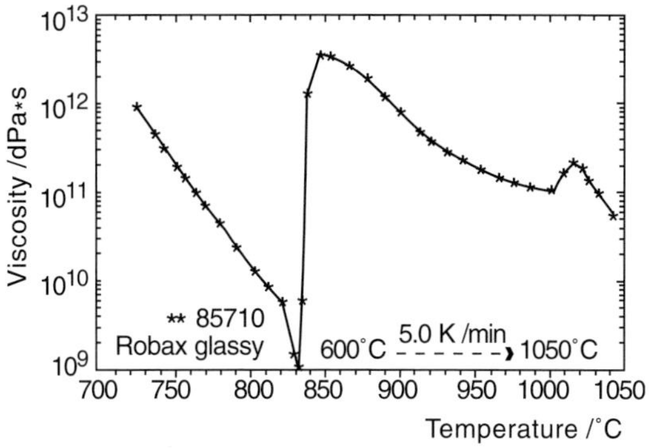

Fig. 2.22. Variation of the viscosity of Robax® base glass during ceramization with a heating rate of 5 K/min

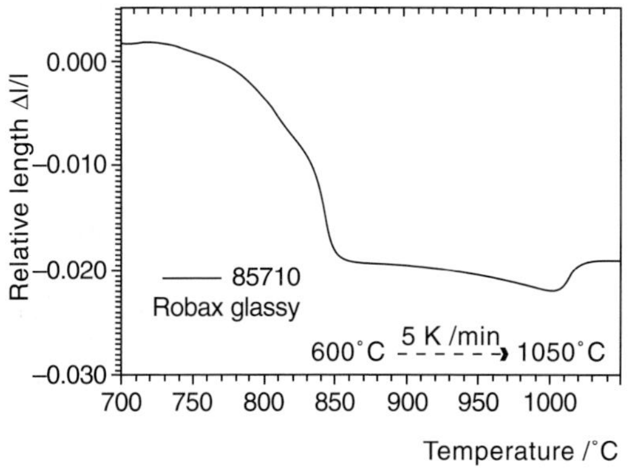

Fig. 2.23. Length variation of Robax® base glass during ceramization with a heating rate of 5 K/min

a large increase in stiffness the ceramization has to be chosen in such a way that the article under consideration is either subjected homogeneously to the time–temperature cycle or free of any external forces; otherwise the article will warp or show other kinds of deformation; striae especially, which ceramize earlier or later than the rest of the article, can cause substantial deformations.

The increase in length at about 1000 °C is evidence for the transformation of the h-quartz s.s. containing glass ceramic into a keatite s.s. containing glass ceramic. This transformation will be discussed at the end of this section.

Although the in-situ measurement of single properties is of interest for monitoring the phase transformation of the base glass into a glass ceramic, another point of considerable interest is the investigation of the development of the microstructure during the phase transformation. This can be partly performed by in-situ experiments using high-temperature X-ray analysis. But this method can give answers to only some of the questions arising as the sensitivity of the method is not sufficient to detect more than the formation of the main crystalline phase, particularly at elevated temperatures. Therefore, most investigations of the development of the microstructure during ceramization have been performed at room temperature on specimens which have been quenched from intermediate stages. The method revealing most insight into the process is transmission electron microscopy. The first concise investigation relevant for commercial compositions has been performed by *Doherty* et al. [2.74] when investigating Corning Code 9608 glass ceramic. Later *Maier* [2.62] obtained rather similar results when investigating Zerodur®. In the following, some of the results of Maier are described. Further information, which in part extends the findings of Doherty et al. and Maier but in other parts deviates significantly from them due to the systems investigated, may be found in [2.69, 75–84].

During heat treatment, Zerodur® base glass phase separates first, see Fig. 2.24, and later small nuclei of about 5 nm in size precipitate, Fig. 2.25. If the temperature is not increased too rapidly, a large number of nuclei are formed before any crystal nucleation and growth of h-quartz s.s. is initiated. If this condition is observed during ceramization, then a high density of small h-quartz s.s. crystals can be obtained. Crystal densities for the h-quartz s.s. crystals are estimated to be about $10^{16}\,cm^{-3}$, whereas the density of the nuclei is not known but may be even higher.

The question of whether the h-quartz s.s. crystals grow epitaxially onto the nuclei has not yet been resolved unambiguously. In the case of the Zerodur® base glass composition there are some strong indications to support this hypothesis. By using a special heat treatment by which only very few nuclei of (Zr, Ti)O_2 type are formed, *Maier* [2.62] was able to produce a microstructure in which a rather large number of h-quartz s.s. crystals have nuclei in their centers. This suggests that the h-quartz s.s. crystals have been nucleated on the surfaces of the (Zr, Ti)O_2 nuclei and then have grown radially to their final size. A direct proof of an epitaxial relationship between both crystal types failed as the intensities of the diffraction pattern of a single (Zr, Ti)O_2 nucleus are too weak.

As has been pointed out already several times in this section, the h-quartz s.s. crystals often form an intermediate metastable phase when glass

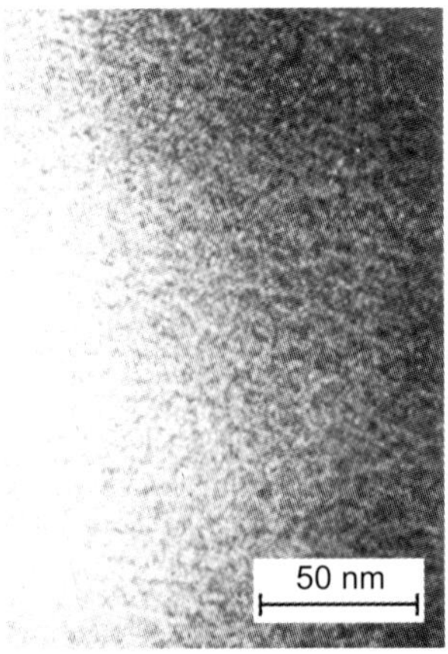

Fig. 2.24. TEM micrograph of heat-treated Zerodur® base glass showing phase separation; heating from 620 °C to 680 °C with a heating rate of 1.67 K/h, then quenched (from [2.62])

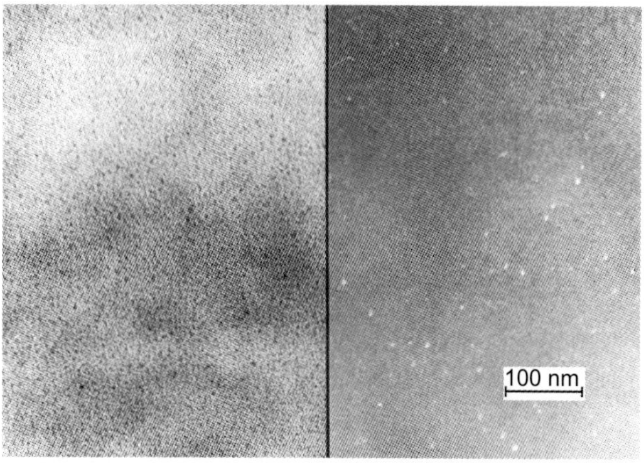

Fig. 2.25. TEM micrograph of heat-treated Zerodur® base glass showing the formation of (Zr, Ti)O_2 nuclei; heating from 620 °C to 700 °C with a heating rate of 1.67 K/h, then quenched. *Left-hand side*: bright field image. *Right-hand side*: dark-field image (from [2.62])

ceramic base glasses are devitrified. At elevated temperatures or prolonged heating times these crystals transform into the more stable keatite s.s. crystals as is evidenced, for example, from Figs. 2.21 and 2.23. What processes initiate this transformation has up to now not been understood. In several cases [2.62, 72, 74] the primary nucleating phase disproportionates into rutile and some less well distinguished phases while the h-quartz s.s. crystals still form the dominant phase.

Another probably important aspect has been pointed out by *Günter* [2.72]. The DTA trace in Fig. 2.23 is not typical of LAS glass ceramics. For many compositions and heat treatment schedules no exothermic peak is observed for the transformation of h-quartz s.s. crystals into keatite s.s. crystals; the observation that this transformation is split into two peaks, and hence into two parts, is even less common. Günter interprets the first peak as being due to recrystallization processes by which small h-quartz s.s. crystals form larger ones. This peak can be observed only when h-quartz s.s. crystals that are small enough have been produced during the glass to h-quartz s.s. transformation. The second peak is due to the h-quartz s.s. to keatite s.s. transformation, which occurs after larger h-quartz s.s. crystals have been formed by recrystallization. The reaction enthalpy of the h-quartz s.s. to keatite s.s. transformation is by more than one order of magnitude smaller than that of the glass to h-quartz s.s. transformation, which explains why the first one is more difficult to observe; the values determined by Günter are 1.9 kJ/mol compared to 27.7 kJ/mol, respectively, for the composition of the Robax® glass ceramic.

The kinetics of the h-quartz s.s. to keatite s.s. transformation has been investigated by *Pannhorst* et al. [2.85] using the Johnson–Mehl–Avrami approach. Later *Strnad* et al. [2.86] presented a similar investigation for another base glass composition. Although the kinetics of the transformation probably depend on many parameters, such investigations help to optimize base glass compositions for applications in which the articles are subjected to high temperatures over prolonged times. In this case compositions have to be chosen for which the initiation of that phase transformation has to be avoided over the lifetime of the article.

References

2.1 C.T. Li: "The crystal structure of $LiAlSi_2O_6III$ (high-quartz solid solution)", Z. Kristallogr. **127**, 327–348 (1968)
2.2 H.G.F. Winkler: "Synthese und Kristallstruktur des Eukryptits, $LiAlSiO_4$", Acta Crystallogr. **1**, 27–34 (1948)
2.3 V. Tscherry, H. Schultz, F. Laves: "Average and superstructure of β-eucryptite ($LiAlSiO_4$) I", Z. Kristallogr. **135**, 161–174 (1972)
2.4 V. Tscherry, H. Schultz, F. Laves: "Average and superstructure of β-eucryptite ($LiAlSiO_4$) II", Z. Kristallogr. **135**, 175–198 (1972)
2.5 W.W. Pillars, D.R. Peacor: "Crystal structure of beta eucryptite as a function of temperature", Am. Mineral. **58**, 681–690 (1973)
2.6 H. Guth: "Strukturuntersuchungen an dem eindimensionalen Li-Ionenleiter β-Eukryptit ($LiAlSiO_4$) mit Hilfe von Neutronenbeugung", KfK-Bericht **2851**, Kernforschungszentrum Karlsruhe (Karlsruhe 1979)
2.7 G. Steinmann: "Strukturuntersuchung an dem eindimensionalen Ionenleiter β-Eukryptit ($LiAlSiO_4$) bei hohen Temperaturen (767 °C) mit Hilfe von Neutronenbeugung", Diploma Thesis (Karlsruhe 1984)
2.8 C.T. Li, D.R. Peacor: "The crystal structure of $LiAlSi_2O_6II$ ("β-spodumene")", Z. Kristallogr. **126**, 46–65 (1968)
2.9 C.T. Li: "Transformation mechanism between high-quartz and keatite phases of $LiAlSi_2O_6$ composition", Acta Crystallogr. B **27**, 1132–1140 (1971)
2.10 F. Liebau: "Einteilung und Mechanismen von Phasenumwandlungen", Fortschr. Mineral. **61**, 29–84 (1983)
2.11 H. Schulz, W. Hoffmann, G.M. Muchow: "Average structure of $Mg[AlSi_3O_{10}]$, a stuffed derivative of the high-quartz structure", Z. Kristallogr. **134**, 1–27 (1971)
2.12 M. Behruzi, T. Hahn: "Struktur und thermische Ausdehnung des ungeordneten β-Eukryptits", Fortschr. Mineral. **55**, Beiheft 1, 12–13 (1977)
2.13 E.M. Levin, C.R. Robbins, H.F. McMurdie (Eds.): *Phase Diagrams for Ceramists* (Am. Ceram. Soc., Columbus 1968) p. 168
2.14 G.H. Beall, B.R. Karstetter, H.L. Rittler: "Crystallization and chemical strengthening of stuffed β-quartz glass ceramics", J. Am. Ceram. Soc. **50**, 181–190 (1967)
2.15 W. Schreyer, J.F. Schairer: "Metastable solid solutions with quartz-type structures on the join SiO_2-$MgAl_2O_4$", Z. Kristallogr. **116**, 60–82 (1961)
2.16 J. Petzoldt: "Metastabile Mischkristalle mit Quarzstruktur im Oxidsystem Li_2O-MgO-ZnO-Al_2O_3-SiO_2", Glastechn. Ber. **40**, 385–396 (1967)
2.17 S. Ray, G.M. Muchow: "High-quartz solid solution phases from thermally crystallizend glasses of compositions (Li_2O, MgO)$\cdot$ $Al_2O_3\cdot nSiO_2$", J. Am. Ceram. Soc. **51**, 678–682 (1968)

2.18 G.H. Beall, D.A. Duke: "Transparent glass-ceramics", J. Mater. Sci. **4**, 340–352 (1969)
2.19 A.J. Perrotta, R.O. Savage: "Beta eucryptite crystalline solutions involving p5+", J. Am. Ceram. Soc. **50**, 112 (1967)
2.20 J. Petzoldt: "Der Einbau von P_2O_5 in metastabile Mischkristalle mit Quarzstruktur des Grundsystems Li_2O-MgO-ZnO-Al_2O_3-SiO_2", Glastechn. Ber. **41**, 181–189 (1968)
2.21 S. Ray: "Solid solutions in the keatite crystal lattice", J. Am. Ceram. Soc. **54**, 213–215 (1971)
2.22 M. Behruzi, T. Hahn: "Hoch-$LiAlSiO_4$ und verwandte Phasen im System $LiAlSiO_4$-$LiAlGeO_4$-$LiGaGeO_4$", Z. Kristallogr. **133**, 405–421 (1971)
2.23 B. Baumgartner, G. Müller: "Framework distortions by large ions in $MAlSi_2O_6$ alumino-silicates with keatite structure", Europ. J. Mineral. **2**, 155–162 (1990)
2.24 G. Müller, H. Paulus, J. Stiefel: "Synthesis and structure of β-quartz type $Na_{0.5}H_{0.5}AlSi_2O_6$ compared to $LiAlSi_2O_6$", N. Jb. Miner. Mh., 493–503 (1990)
2.25 G. Müller, M. Hoffmann, R. Neeff: "Hydrogen substitution in lithium-alumino-silicates", J. Mater. Sci. **23**,1779–1785 (1988)
2.26 M. Sternitzke, G. Müller: "Substitutional and thermal expansion in $MAlSi_2O_6$ aluminosilicates with keatite structure", Eur. J. Mineral. **3**, 769–776 (1991)
2.27 F.H. Gillery, E.A. Bush: "Thermal contraction of β-eucryptite ($Li_2O \cdot Al_2O_3 \cdot 2SiO_2$) by X-ray and dilatometer methods", J. Am. Ceram. Soc. **42**, 175–177 (1959)
2.28 J.S. Moya, A.G. Verduch, M. Hortal: "Thermal expansion of β-eucryptite solid solutions", Trans. Brit. Ceram. Soc. **76**, 177–178 (1974)
2.29 M. Hortal, R. Villar, S. Vieira, J.S. Moya: "Linear isothermal compressibilities of β-eucryptite", J. Am. Ceram. Soc. **58**, 262 (1975)
2.30 H. Schulz: "Thermal expansion of β-eucryptite", J. Am. Ceram. Soc. **57**, 313–326 (1974)
2.31 U. v. Alpen, E. Schönherr, H. Schulz, G.H. Talat: "β-eucryptite, a one-dimensional Li-ionic conductor", Electrochim. Acta **22**, 805–807 (1977)
2.32 R.A. Young: *Mechanism of the Phase Transition in Quartz* (Georgia Institute of Technology, Atlanta 1962) AFOSR-2569
2.33 R.M. Hazen, L.E. Finger: *Comparative Crystal Chemistry* (Wiley, Chichester 1982)
2.34 H.T. Smyth: "The role of transverse oxygen vibrations in thermal expansion behaviour of glasses and crystals", in *Thermal Expansion*, ed. by M.G. Graham, E.E. Hagy (Am. Inst. of Physics, New York 1971)
2.35 H.D. Megaw: *Crystal Structures, a Working Approach* (Saunders, Philadelphia 1973)
2.36 R.J. Ackermann, C.A. Sorrell: "Thermal expansion and the high-low transformation in quartz I: High temperature X-ray studies", J. Appl. Crystallogr. **7**, 461–467 (1974)
2.37 P.P. Keat: "A new crystalline silica", Science **120**, 328–330 (1954)
2.38 G. Müller, M. Sternitzke: "Computer modelling of structure and thermal expansion of β-quartz- and keatite-type alumino-silicates", J. Mater. Sci. Lett. **12**, 278–280 (1993)
2.39 M. Sternitzke, G. Müller: "Crystal structure and thermal expansion of quartz-type aluminosilicates", J. Mater. Sci. **26**, 3051–3056 (1991)
2.40 J.W. Christian: *The Theory of Transformations in Metals and Alloys*, Pt. 1, 2nd ed. (Pergamon, Oxford 1975)
2.41 S. Toshev, I. Gutzow: "Time lag in heterogeneous nucleation due to non-stationary effects", Phys. Status Solidi **21**(2) 683–691 (1967)

2.42 U. Schiffner: "Keimbildung und Kristallwachstum in Ausgangsgläsern handelsüblicher Li_2O-Al_2O_3-SiO_2-Glaskeramiken", PhD Thesis (Erlangen-Nürnberg 1984)
2.43 U. Schiffner, W. Pannhorst: "Nucleation in a precursor glass for a Li_2O-Al_2O_3-SiO_2-glass ceramic, Part 1. Nucleation kinetics", Glastechn. Ber. **60**, 211–221 (1987)
2.44 U. Schiffner, W. Pannhorst: "Nucleation in a precursor glass for a Li_2O-Al_2O_3-SiO_2-glass ceramic, Part 2. Variation of the nucleating agent concentrations", Glastechn. Ber. **60**, 239–247 (1987)
2.45 G. Tammann: *Kristallisieren und Schmelzen* (Barth, Leipzig 1903)
2.46 G. Tammann: *Aggregatszustände*, 2nd ed. (Voss, Leipzig 1923)
2.47 M. Volmer, A. Weber: "Keimbildung in übersättigten Gebilden", Z. Phys. Chemie **119**, 227 (1926)
2.48 R. Becker, W. Döring: "Kinetische Behandlung der Keimbildung in übersättigten Dämpfen", Ann. Physik **24**, 719–752 (1935)
2.49 D. Turnbull, J.C. Fischer: "Nucleation rate in condensed systems", J. Chem. Phys. **17**, 71 (1949)
2.50 J.B. Zeldovich: "On the theory of new phase formation", Acta Physicochim. URSS **18**, 1–22 (1943)
2.51 J. Frenkel: *Kinetic Theory of Liquids* (Oxford University Press, Oxford 1946)
2.52 D. Kashchiev: "Solution of the non-steady state problem in nucleation kinetics", Surf. Sci. **14**, 209–220 (1969)
2.53 S.D. Stookey: "Catalyzed crystallization of glass in theory and practice", in *Internationaler Glaskongreß*, Sonderband V, Glastechn. Ber. **32K** (V) (1959)
2.54 P.W. McMillan: *Glass-Ceramics*, 2nd ed. (Academic Press, London 1979)
2.55 W. Sack, H. Scheidler: "Einflußder Keimbildner TiO_2 und ZrO_2 auf die sich ausscheidenden Kristallphasen bei der Bildung von Glaskeramik", Glastechn. Ber. **39**, 126–130 (1966)
2.56 D.R. Stewart: "TiO_2 und ZrO_2 as nucleants in a lithia aluminosilicate glass-ceramic", in *Advances in Nucleating and Crystallization in Glasses*, ed. by L.L. Hench, S.W. Freiman (Am. Ceram. Soc., Columbus 1971)
2.57 G. Müller: "Zur Wirkungsweise von Gemischen oxidischer Keimbildner in Glaskeramik des Hochquarz-Mischkristalltyps", Glastechn. Ber. **45**, 189–194 (1972)
2.58 R.D. Maurer: "Crystal nucleation in a glass containing titania", J. Appl. Phys. **33**, 2132 (1962)
2.59 G.H. Beall: B.R. Karstetter, H.L. Rittler: "Crystallisation and chemical strengthening of stuffed β-quartz glass ceramics", J. Am. Ceram. Soc. **50**, 181–190 (1967)
2.60 P.E. Doherty, D.W. Lee, R.S. Davis: "Direct observation of the crystallization of Li_2O-Al_2O_3-SiO_2-glasses containing TiO_2", J. Am. Ceram. Soc. **50**, 77–80 (1967)
2.61 W. Vogel: *Struktur und Kristallisation der Gläser* (VEB Deutscher Verlag für Grundstoffindustrie, Leipzig 1965)
2.62 V.M. Maier: "Untersuchungen zur Keimbildung und Kristallwachstum in Alumosilikat-Glaskeramiken mittels Elektronenmikroskopie und Röntgenabsorptionsspektroskopie", PhD Thesis (Darmstadt 1987)
2.63 W.A. Zdaniewski: "Microstructure and kinetics of crystallization of MgO-Al_2O_3-SiO_2-glass-ceramics", J. Am. Ceram. Soc. **61**, 199–204 (1978)
2.64 T.J. Barry, L.A. Lay, R.P. Miller: "Nucleation efficiency in lithia-alumina-silica glasses", Discuss. Farad. Soc. **50**, 214–221 (1970)
2.65 M.H. Lewis, J. Metcalf-Johansen, P.S. Bell: "Crystallisation mechanisms in glass ceramics", J. Am. Soc. **62**, 278–288 (1979)

2.66 P.F. James: "Nucleation in glass-forming systems – a review. Nucleation and crystallization in glasses", in *Advances in Ceramics*, Vol. 4 (Am. Ceram. Soc., Westerville 1982)
2.67 H. Scheidler, W. Kiefer: personal communication, Schott Glaswerke, 1969
2.68 A.H. Dietzel: *Emaillierung* (Springer, Berlin, Heidelberg 1981)
2.69 G.H. Beall, D.A. Duke: "Transparent glass-ceramics", J. Mater. Sci. **4**, 340–352 (1969)
2.70 E.J. Smoke: "Ceramic compositions having negative linear thermal expansion", J. Am. Ceram. Soc. **34**, 235–239 (1951)
2.71 J. Petzoldt: "Metastabile Mischkristalle mit Quarzstruktur im Oxidsystem Li_2O-MgO-ZnO-Al_2O_3-SiO_2", Glastechn. Ber. **40**, 385–396 (1967)
2.72 C. Günter: "Untersuchungen der kristallchemischen Ursachen für die Umwandlung der Hochquarz- in Keatit-Mischkristalle in Li_2O-$Al_2O_3SiO_2$-Glaskeramiken", PhD Thesis (Darmstadt 1992)
2.73 H. Scheidler, W. Sack: "Die unterschiedliche Wirkung der Keimbildner ZrO_2 und TiO_2 auf das Kristallisationsverhalten eines Lithium-Aluminium-Silikat (Glaskeramik)-Glases", Proc. 9th Int. Congr. of Glass, Versailles, France, Sept. 27–Oct. 2, 1971, pp. 1069–1085
2.74 P.E. Doherty, D.W. Lee, R.S. Davis: "Direct observation of the crystallization of Li_2O-Al_2O_3-SiO_2-glasses containing TiO_2", J. Am. Ceram. Soc. **50**, 77–80 (1967)
2.75 W. Sack: "Glas, Glaskeramik und Sinterglaskeramik", Chemie-Ing. Techn. **37**, 1154–1165 (1965)
2.76 T.I. Barry, D. Clinton, L.A. Lay, R.A. Mercer, R.P. Miller: "The crystallization of glasses based on eutectic compositions in the system Li_2O-Al_2O_3-SiO_2; part 1: Lithium metasilicate – β-spodumene", J. Mater. Sci. **4**, 596–612 (1969)
2.77 T.I. Barry, D. Clinton, L.A. Lay, R.A. Mercer, R.P. Miller: "The crystallization of glasses based on eutectic compositions in the systems Li_2O-Al_2O_3-SiO_2; part 2: Lithium metasilicate – β-eucryptite", J. Mater. Sci. **5**, 117–126 (1970)
2.78 T.I. Barry. L.A. Lay, R.P. Miller: "Nucleation efficiency in lithia-alumina-silica glasses", Discuss. Farad. Soc. **50**, 214–221 (1970)
2.79 G.H. Beall: "Structure, properties, and application of glass-ceramics", in *Advances in Nucleation and Crystallization in Glasses*, Vol. 5 (Am. Ceram. Soc., Columbus 1971) pp. 251–261
2.80 M.L. Wang, R. Stevens, P. Knott: "Microstructure and crystallization behaviour of a transparent glass ceramic doped with Fe_2O_3", Glass Techn. **23**, 238–243 (1982)
2.81 B. Andrianasolo, B. Champagnon, C. Esnouf: "Ultrafine grained glass-ceramics obtained with Cr_2O_3 additions", J. Non. Cryst. Solids **126**, 103–110 (1990)
2.82 J.Y. Hsu, R.F. Speyer: "Comparison of the effects of titania and tantalum oxide nucleating agents on the crystallization of $Li_2O{\cdot}Al_2O_3{\cdot}6SiO_2$ glasses", J. Am. Ceram. Soc. **72**, 2334–2341 (1989)
2.83 J.Y. Hsu, R.F. Speyer: "Influences of zirconia and silicon nucleating agents on the devitrification of $Li_2O{\cdot}Al_2O_3{\cdot}6SiO_2$ glasses", J. Am. Ceram. Soc. **73**, 3585–3593 (1990)
2.84 J.Y. Hsu, R.F. Speyer: "Crystallization of $Li_2O{\cdot}Al_2O_3{\cdot}6SiO_2$ glasses containing niobium pentoxide as nucleating dopant", J. Am. Ceram. Soc. **74**, 395–399 (1991)
2.85 W. Pannhorst, W. Wichelhaus: "Untersuchungen zur Umwandlung von h-Quarz-Mischkristallen in Keatit-Mischkristalle in Li_2O-Al_2O_3-SiO_2 Glaskeramiken", Glastechn. Ber. **56K**, 572–577 (1983)

2.86 Z. Strnad, M. Wada: "Volume and surface crystallization during formation of glass-ceramic materials", in *Proc. 15th Int. Congr. on Glass*, Vol. 3b (Leningrad 1989) pp. 106–110

3. Glass Ceramics for Household Appliances

3.1 Cooking Systems with Ceran®: High-Tech Appliances for the Kitchen

Eva Willhauk, Ruban Harikantha

Since the beginning of serial production of in January 1973 more than 60 million Ceran® cooktop panels have been sold around the world. A market share of more than 70% for glass ceramics in sales of electric cooking appliances in Western Europe underlines the success of this product as an effective symbiosis of aesthetics and perfect technique. The properties of this exceptional material have set new standards in modern kitchens and revolutionized cooking habits in Europe.

The very first inventions using the new material glass ceramic were made by Stookey in the 1950s. He developed a strong and thermal shock resistant white aluminosilicate glass ceramic called Pyroceram® 9608 (Corning Glass Works, USA). This material was the first to be marketed for use as household crockery; it was also called Corning Ware® 9608 cookware in 1959 [3.1]. Another commercial product made of this material was rocket nose cones.

In the late 1960s a new type of electric cooker [3.2] with white Pyroceram® as cooktop panels was marketed in the USA. Meanwhile opaque white glass-ceramic surfaces in the USA were not only sold by Corning Glass Works, but also by PPG (Hercuvit™) and Owens-Illinois (Cer-Vit®). All three were processed differently. Pyroceram® was a rolled and polished glass ceramic, chemically hardened to enhance strength. Hercuvit™ was also a rolled material, with higher infrared transmission than Pyroceram®, but aesthetically less attractive. Cer-Vit® was pressed in tiles having the dimensions of a single-sized hotplate; so a few of these tiles were required to build a cooking surface in a metal frame. The ease of cleaning of this solution was more than doubtful.

Also in Japan in 1962, a white nontransparent β-spodumene solid-solution glass ceramic was produced. Nippon Electric Glass manufactured cooktop panels under the trademark Neoceram™ N-11. With a coefficient of thermal expansion of approximately $1.3 \times 10^{-6}\mathrm{K}^{-1}$, they were used for induction cooker top plates, kitchenware, and optical components.

Another important and economically significant β-quartz solid solution glass ceramic for domestic appliances was Keraglass® from Eurokera (joint venture of Corning/St. Gobain, USA/France), introduced into the market as EuroKera® cooktop panels at the end of 1993.

In the beginning all manufacturers along the entire value chain lacked the competence to implement a glass ceramic into a cooking system. This was a big obstacle to successful market penetration. Therefore Schott decided to establish a "technical application laboratory" in order to develop, together with customers and suppliers, an entire cooking system. Besides the glass ceramic panel, these developments also included electric heating devices, framing, assembly, and cookware. By these efforts Schott achieved its goal to create a new black cooktop panel on the basis of a β-quartz solid solution glass ceramic, called Ceran®, with the desired and above-mentioned market penetration.

The success of Ceran® is partly also due to the experience and know-how gained in the development and production of Zerodur®. Schott already started producing this technical glass ceramic in the 1960s. With its exceptional temperature stability (non-deforming when subjected to variations between $-20\,°C$ and $+100\,°C$) Zerodur® has become the material of choice for optical precision applications such as space telescopes, for example (see Chap. 4).

Its virtually zero thermal expansion coefficient made this material interesting for other applications, too. The idea of using Zerodur® for cookware soon came to mind. Trademarks, such as Pyroflam®, Jena 2000®, Ceradur® and Corning Vision® produced by the main suppliers Corning Glassworks and Schott, are well-known cookware articles. At the end of the 1960s this product benefit led to the development of Ceran® cooktop panels.

Whether subjected to different high temperatures over various lengths of time or rapid and repeated switching on and off of the cooking temperature, the shape and size of the cooking surface remain unchanged. Alongside this Ceran® transfers large amounts of heat to the pan while conducting very little heat sideways across the cooking surface. As a result the unheated zones remain cold and can be touched without risk. In addition the material is highly resistant to breakage when subjected to all usual mechanical loads.

The first Ceran® cooktop panels had their market launch in 1972 and, after making slow progress at first, conquered the markets in Germany and other countries more and more rapidly. From the very start glass ceramic cooktop panels were looked on as components for an up-to-date cooking system and embedded in a comprehensive system philosophy. Instead of seeing the cooktop panel as an isolated product, Schott has handled it as a core element of a complete cooking system: other components – from the heating element, control unit, cookware plus the frame and enclosure right up to cleaners – are also included and continuously optimized.

A determining factor for this approach was the realization that optimization can only be achieved if all the parameters are seen as a unit. This insight provided the basis for the Ceran-Top-System® (CTS). It thus represents an expansion of the service provided by Schott in the product environment of Ceran® cooktop panels, identifies components that influence the performance of the cooking system, names evaluation criteria, and gives recommendations for an optimum cooking system. In this way a range, rich in variants, came into being, whose optimization will be continued in dialog with everyone concerned.

Schott linked this system thinking with a new concept, "Integrated Marketing", which, in addition to the technological integration of the partners, also planned the investigation of the needs and requirements at purchaser levels as well as appropriate communication measures.

Ceran® owners in the major Western European countries are united in naming ease of cleaning and care of the flat cooking surface as the main reason for the purchase of a glass-ceramic cooktop; followed by its attractive and aesthetic appearance plus its modern and practical handling features. User surveys in America and other countries produce similar results.

In 1984 Schott received the renowned "Deutscher Marketing-Preis" ("German Marketing Prize") in recognition of the new marketing approach by which Schott, originally a production goods manufacturer and remote from the market, was transformed into a market-oriented company that "has developed new markets and changed the structure of its product range".

By 1977 all the well-known European cooking appliance manufacturers had Ceran® cooktop panels in their product range. The one millionth cooktop panel was shipped in 1980; the ten-millionth was made in 1989; the twenty million mark was reached in 1993; and in 2002 the total was already over fifty million.

Ten years after its market launch some 35% of all cooking appliances purchased each year in Germany were fitted with glass ceramic. Nowadays more than 50% of all households in Germany cook on a range with Ceran® cooktop panels – that is around eighteen million satisfied households. This successful development very soon spread to other markets as well (Fig. 3.1). In Japan, an Asian core market, the yearly market volume of Ceran® cooktop panels averages about five and a half million units.

After some initial difficulties Schott also managed to convince cooking appliance manufacturers in the USA of the efficiency of the Ceran-Top-System®. In reaction to the continuous rise in Ceran® cooktop panel exports to North America, the post-processing of Ceran® cooktop panels for the North American market was transferred to Vincennes, Indiana, in 1991.

As a result of the consistent implementation of the system philosophy the Ceran® cooktop panel has also breathed life into the further development of all other components. An example of this is the new type of multi-ring heating element, which can be ideally adapted to various pan sizes while sav-

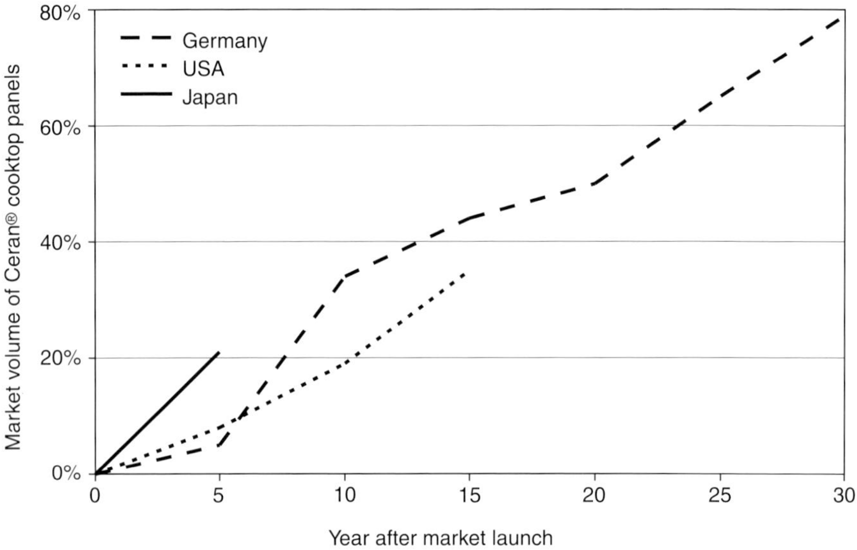

Fig. 3.1. Development of the market volume of Ceran® glass-ceramic cooktops in Germany, USA, and Japan after market launch

ing energy. Any heating system using electricity or gas can now be combined with glass ceramic cooktop panels to achieve the shortest possible boil-up time with a low energy requirement.

The latest innovation is the further development of the material to produce the new Ceran Suprema® glass-ceramic cooktop panel. Optimization of the composition of the material plus a new manufacturing process resulted in a further improvement of the heat transmission and temperature resistance of the glass ceramic. In consequence, built-in radiant heating elements can now be adjusted to higher cooking zone temperatures. And boil-up time, depending on the cookware used, can be as much as 20% less (see also Sect. 3.2.3).

There have been gas cooktops with glass ceramics for many years. An example of an adaptation to suit a particular market is provided by the close cooperation between Schott and a cooking appliance manufacturer on a new gas burner system (Inner Burner) in Japan. As a way of opening up the Japanese market this innovative concept for a gas cooking appliance incorporating a Ceran® cooktop panel was implemented in a consistent way taking into account the cooking habits specific to the market. It was the high temperature resistance of the glass ceramic in particular that made it possible to combine the high-power burner directly with the cooktop panel.

Innovations have been bringing new improvements all the time. The design variety for product differentiation is manifold. Be it unusual new different shapes or edge and frame designs (Fig. 3.2a), system features such as touch controls and heating functions (Fig. 3.2b) or a new exciting selection of suba)

b)

Fig. 3.2. Ceran® offers a variety of forms (**a**) and functions (**b**)

strate and decoration colors, there are virtually no limits when it comes to creating a "new look".

While in the last thirty years only black cooktop panels (Fig. 3.3 left) have been marketed, recently white and cream cooktop panels have been introduced. This development responds to the request for more differentiation. As a result, Schott has started a campaign to introduce a number of new cooktop panels which are colored in the bulk. White and cream-colored cooktop panels are based on keatite s.s. crystal containing glass ceramics (see the Ceran® Arcticfire™ cooktop panel in Fig. 3.3 right). The variety of colors available to date can be embellished with a broad selection of decoration colors (see Fig. 3.4).

In noted shows and fairs such as Home Tech (a prominent European appliance show) and NKBIS (New Kitchen and Bath International Show) Schott has introduced Ceran Cook-In® and the Cook-Box™ (Fig. 3.5) as design studies to demonstrate the vast range of design possibilities such as shapes, displays, heatings, and lights offered by the unique properties of "construction" materials such as glass and glass ceramic.

Today Schott offers cooktop panels in combination with other components. The objective in this is to remain true to the company's core competence but at the same time to offer the customer added value. Thus, instead of individual components and services, assemblies right up to complete systems can be supplied. With its "one-stop shopping" strategy Schott relieves its customers of work-intensive production stages and responds to their demand for cost saving.

Innovative added value solutions, for example touch control systems, have been developed in close partnership with appliance manufacturers and component producers. These offer our customers the necessary flexibility in their

Fig. 3.3. Black Ceran® glass ceramic cooktop panel (*left*) and white Ceran® Arcticfire™ glass ceramic cooktop panel (*right*)

Fig. 3.4. Selection of decoration colors for Ceran®

design concepts with new electronic touch-sensitive surfaces. Additionally they make it possible to produce previously unavailable control units for the intelligent control of cooktops. In this way, for example, separate cooking zones can be configured individually on a single- or multi-circuit basis and boil-up phases of individual cooking zones can be programmed variably. In so doing Schott brings together various competences to a complete solution and uses the electronics know-how of an affiliated company.

Schott has also set itself ambitious goals for the future. In grouping together Schott's wide-ranging competence in close cooperation with its customers, strengthening their position in the market is always the central focus of its dealings. In this way Schott reacts with appropriate products to consumer and lifestyle trends and in so doing attempts to support its customers' product differentiation.

The ongoing trend of domestic sociability has impacted the design of kitchens: kitchen-living rooms are again the in-thing. But there are no obvious color and material trends dominating all the rest; the wide variety of kitchen

Fig. 3.5. Cook-Box^{TM} – a design study to present the manifold creative possibilities of glass and glass ceramic

styles on the market is increasing. Striking products are, therefore, gaining importance.

Convenience and shortage of time are also determining factors in today's lifestyle. The wide variety of uses (fast food, slow food, non-food) for the kitchen is becoming greater. With Ceran Suprema® Schott takes up this trend and provides a new generation of high-performance cooktop panels. Yet the development is by no means over. Through continuous research and close cooperation with component manufacturers and the household appliance industry Ceran® will continue to be a convincing example of innovation in the kitchen in the future.

3.2 Cooking Systems with Ceran®: How it Works

Roland Dudek, Cora Krause, Klaus Kristen, Peter Naß, Kurt Schaupert, Herwig Scheidler, Wolfgang Schmidbauer, Patrik Schober, Martin Taplan, Ted Wegert, Evelin Weiss

As has been noted in Sect. 3.1, market research studies are periodically carried out to learn more about explicit customer demands as well as their favored product features. For years the polls have shown the following functions and properties to play a predominant role:

- reduced boil-up times,
- sensitive monitoring and control during boil-up and cooking action in order to avoid a boil-over,
- easy-to-clean and scratch-resistant surface,
- an appealing design, for example by new shapes or edge designs,
- attractive high variety of cooktop (bulk) materials and decoration colors.

In daily use, the cooktop panels are exposed to the following conditions: temperatures up to 750 °C, chemical attack by boiling over and cleansers, and mechanical loads (abrasion) caused by cleansers and moving of saucepans as well as by static and dynamic stresses. A Ceran® cooktop panel with a thickness of only 4 mm withstands all these practical loads that occur in a kitchen. The entire system (consisting of cooktop panel, heating elements with their control devices, housing and framing) fulfills all safety regulations and standards (e.g. "Deutsche Industrienorm" (DIN), "Verein deutscher Elektroingenieure" (VDE), "Underwriters Laboratories" (UL)) and the cooktop retains its attractive appearance, as long as the specified cleaning and maintenance procedures are followed.

3.2.1 The Concept of the Ceran-Top-System®

The concept of the Ceran-Top-System® (CTS, Sect. 3.1) implies that components must not be viewed separately but as a system of components with mutual interaction. Schott works closely together with component suppliers and appliance manufacturers to obtain the optimum appliance for a satisfied consumer. Figure 3.6 illustrates the essential parts of a cooking system.

3.2.2 Heat Flow

Most important for the technical performance of the cooking system – and thus for consumer satisfaction – is the *heat flow* as it influences the boiling time, energy consumption, and temperature distribution in the cookware used.

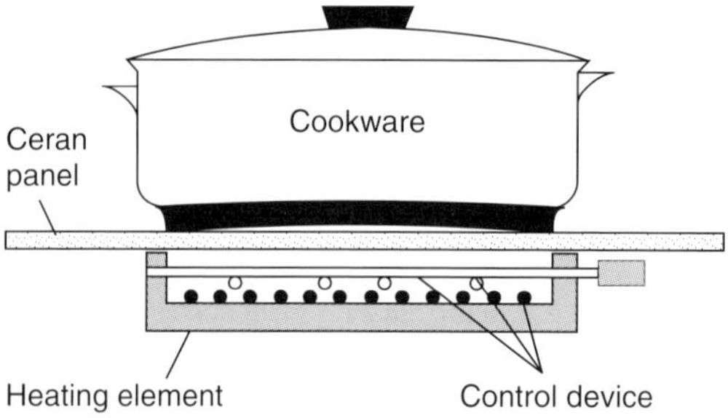

Fig. 3.6. Essential elements of a cooking system

In this section we describe the influence of the various components on the heat flow and give recommendations for their layout. An overview of the energy flow in the cooking system is given in Fig. 3.7. The electrically heated filaments give off their energy into the environment largely as radiant energy (1–4). A part of this energy is passed on to the glass ceramic, where it is either absorbed (1) or, due to transmission, is allowed to pass through (2). The glass ceramic heats up by (1) and transfers this energy directly to the cookware bottom. The energy transfer occurs here mainly through direct contact (8) and only partially through the radiant heat of the glass ceramic (5).

The remainder of the radiant energy from the heating coils reaches the insulation surface where it is mostly reflected (3). Less than 5% is absorbed (4). As the heating coils are in direct contact with the insulation, a small portion of energy is released through direct contact into the insulation (10).

The heat flow between the panel and the cookware bottom is mainly influenced by the air gap separating them. When flat cookware is used with

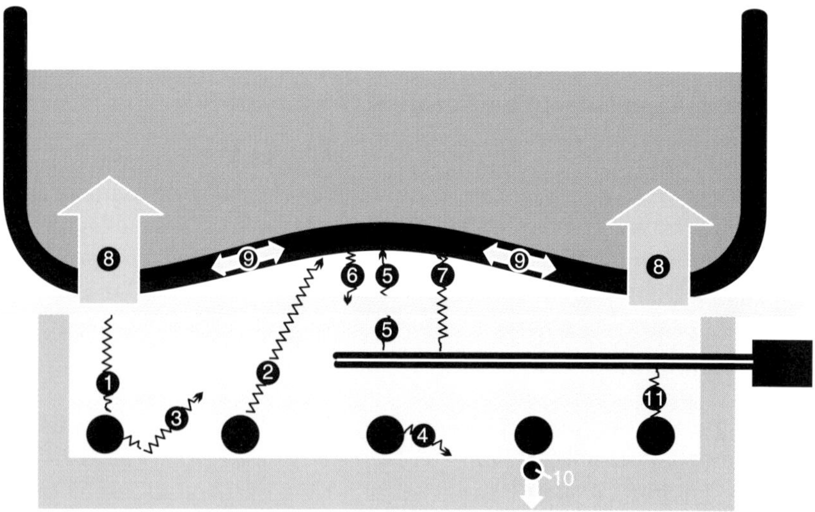

Fig. 3.7. Heat flow in the cooking system

a small air gap (0.2–0.5 mm), the heat flow by contact (8) is very good. The surface temperature of the panel reaches only approximately 300–400 °C, a range where radiant energy (5) plays no role.

If, in the case of poor-quality cookware, a large air gap (> 1 mm) exists, a noteworthy share of energy will be transported by radiation. Depending on the cookware material, the radiant energy will be more or less absorbed from the cookware bottom. Enamel cookware absorbs more than 95%, stainless steel and aluminum cookware nearly 50% of the radiated energy. In any case, the heat flow will be hindered by the air gap, so that, for example, the short boiling time of good-quality cookware (narrow air gap) cannot be attained.

The energy not transferable through conduction (8) must be carried as radiation (2, 5); therefore, the glass ceramic heats up to a higher temperature (e.g., up to 650 °C) than with good cookware. The result is that the temperature limiter will be struck with radiant energy (5, 7). In accordance with its purpose (the protection of the overall system against over-heating) the temperature limiter switches the heating coils on and off. The average heat flow is consequently reduced and the boiling time is increased. Due to the lower reflection (7) of enamel cookware, the increase in the boiling time is less pronounced compared to stainless steel cookware for comparable bottom geometries.

In addition to the quality of cookware, another essential factor for good heat flow is the match of the diameters of heating element and cookware. The quantitative effect of the good surface match is shown in Fig. 3.8. Two qualities of cookware were used, both having the same base diameter (220 mm). The bars in the back represent the test results for poor cookware and the bars in front those for good cookware. Tests measuring the boiling time were conducted on heating elements of 180 mm in diameter (shown on the left side of Fig. 3.8). In this case, the heating element is significantly smaller than the base of the pot. In other words, the base of the poor-quality pot is resting on the cold surface area of the panel, having no contact with the active cooking zone. The result is a boiling time of 20 min.

When the poor-quality pot is placed on a heating element of the same size, the boiling time reduces from 20 to 13.6 min. This means that without an improvement in the quality of the pot, a reduction of 30% is attained through the proper match of the cookware and heating element diameters. Should this poor-quality cookware be placed onto an even larger heating element with a diameter of 230 mm, a further reduction of approximately 15% will result. An increase in the energy output through this heating element has, however, a negative effect. The boiling time increases again as the temperature limiter switches off the surplus energy.

When a good pot is used, characterized by its smooth contact to the glass ceramic surface, one hardly notices the influence from the size of the heating element. Using a large pot on a smaller heating element results in a boiling time of 9.5 min.

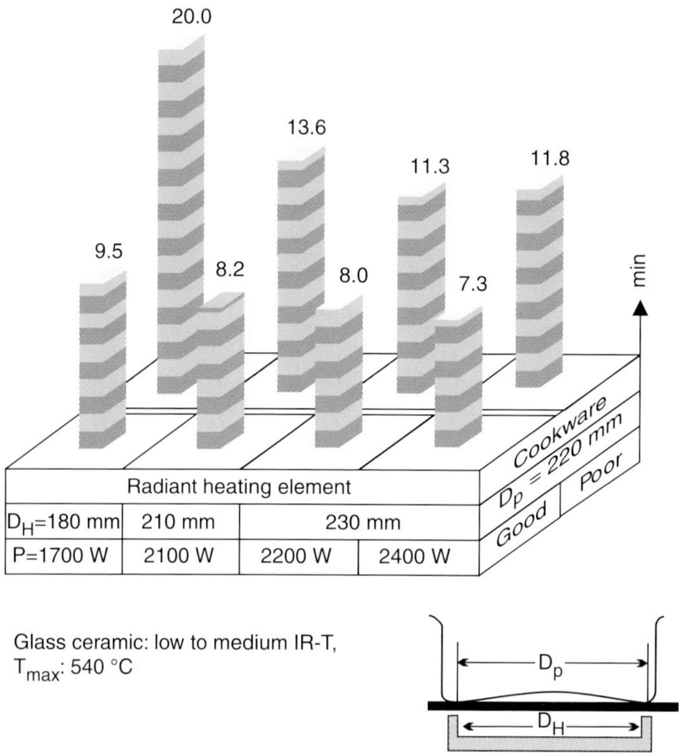

Fig. 3.8. Influence of heating element diameter/wattage and cookware diameter/quality on the heating-up time (for heating 2 l water from 15 to 90 °C)

By matching up cookware to a heating element of the same size, as in the case with the heating element of 210 mm in diameter, we observe a boiling time of 8.2 min. A further increase in the size of the heating element does not lead to a further reduction in the boiling time. The differences shown here are more likely attributable to the different power consumptions of the heating elements.

Besides the right choice and placement of cookware, there are further factors that influence boil-up times and therefore the technical performance of the cooking system. These are:

- type of glass ceramic (transmission, T–t loading),
- type of heating element,
- wattage of heating element,
- setting of temperature limiter,
- quantity of substance to be cooked.

3.2.3 Transmission, Temperature–Time Loading

Schott started its cooktop panel production with Ceran® Color (purple lavender tint). Today Schott produces a variety of cooktop panels made of different glass ceramic types, of which the two types:

- Ceran Hightrans® (brown yellowish tint)
- Ceran Suprema® (brown yellowish tint)

are mainly applied to equip standard cooking systems. The transmission curves for these two types of glass ceramics (panel thickness 4 mm) are shown in Fig. 3.9.

For the Ceran Suprema® glass ceramic panel, Schott combined a new production process with a modified glass composition which led to a material with a 10% higher transmission in the IR range. Higher transmission causes lower absorption of radiation energy in the glass ceramic. This can be quantified in case of free radiation, i.e., when the system is heated without pot and without change of limiter setting. Then the surface temperature of Ceran Suprema® is about 5–7 K lower than that of Ceran Hightrans® under identical conditions when using common ribbon heaters. In this case the influence for reduction of boil-up time caused by higher IR transmission is not very significant.

Apart from higher transmission, Ceran Suprema® also shows a significantly higher T–t loading limit, which specifies the maximum permissible temperatures for given load times. At or below this limit, cracking of cooktop panels due to compaction stress is excluded (see also Sect. 3.2.7). The

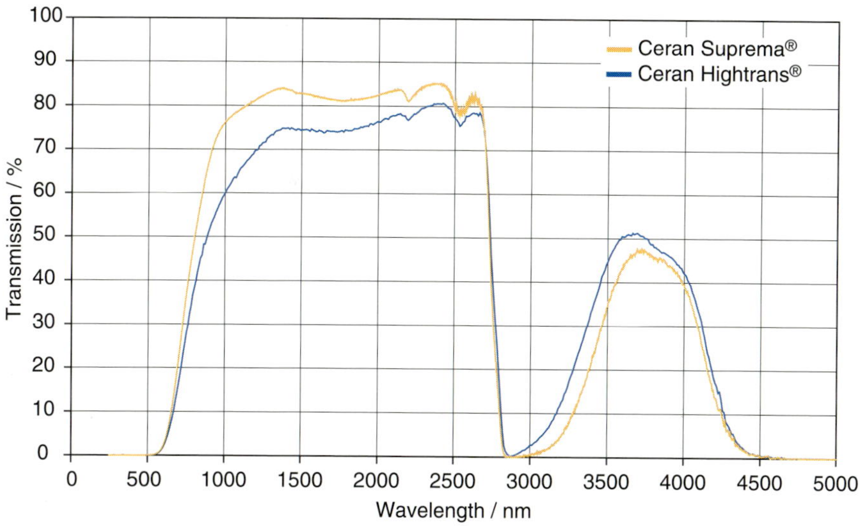

Fig. 3.9. Typical transmission ranges of Ceran Hightrans® and Ceran Suprema® (panel thickness 4 mm)

heating device of the cooking system must be designed to ensure that the T–t loading limit is not exceeded during use. This is commonly achieved by using heating elements with suitable limiter settings (see also Sect. 3.2.2).

In consequence, Ceran Suprema® allows for higher limiter settings of up to 40 K compared to Ceran Hightrans®, resulting in faster boil-up times. The characteristic values are shown in Table 3.1 and Fig. 3.10. By using more powerful heating elements and higher-adjusted settings, boil-up times can be reduced by up to 20%, depending on the cookware used. These findings are shown in Fig. 3.11.

Table 3.1. T–t loading limits for Ceran Hightrans® and Ceran Suprema®

Loading time	Loading temperature limit	
	Ceran Hightrans®	Ceran Suprema®
h	°C	°C
5000	560	600
1000	610	650
100	660	700
10	710	750

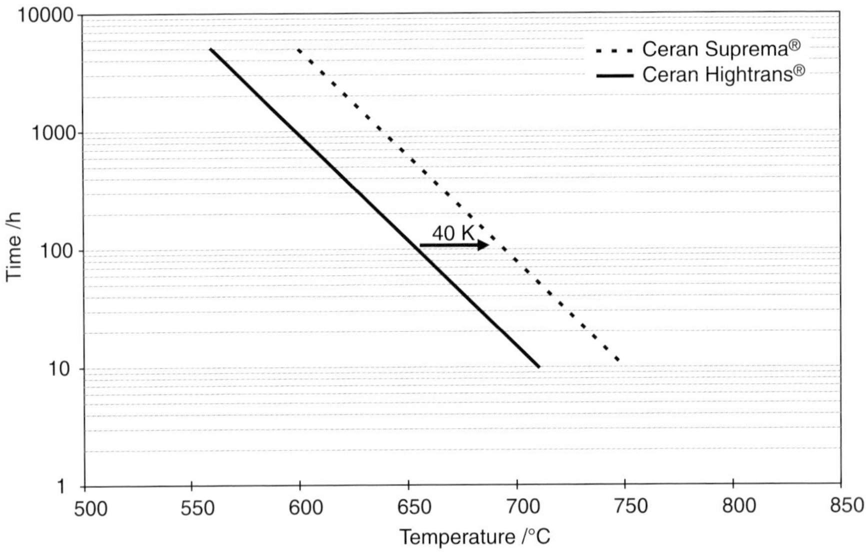

Fig. 3.10. T–t loading ranges for Ceran Hightrans® and Ceran Suprema®

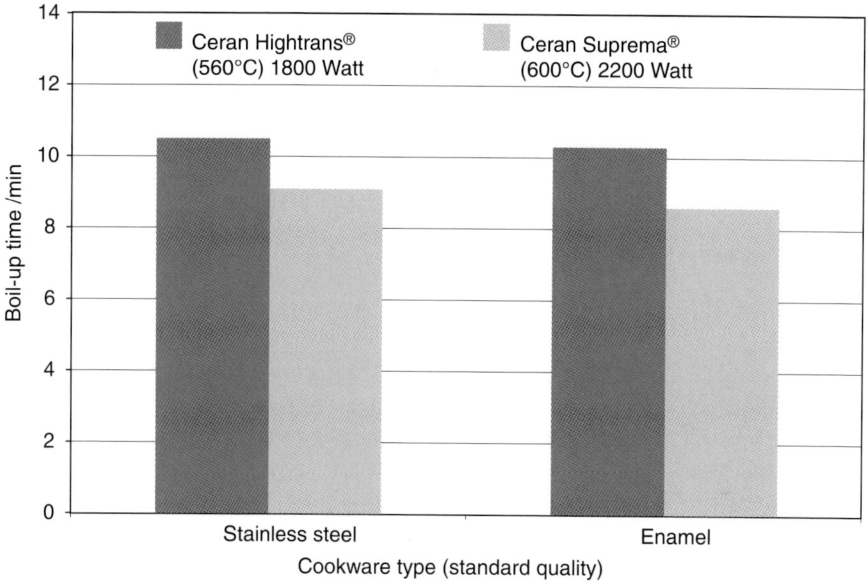

Fig. 3.11. Ceran Hightrans® and Ceran Suprema® in use with standard cookware quality (different cookware materials) and different limiter settings (560°C respectively 600°C)

3.2.4 Heating Elements

The cooking performance is also influenced by the type of heating element. So far, for the greater part, heat flow and energy transfer have been discussed for the most common type of heating element – the *ribbon* heater, which has its radiation peak at 2.5 µm.

Another type of heating element is the *halogen heater* (radiation peak at 1.1 µm), which has a shorter reaction time during the heat-up phase. This heater type is disappearing more and more from the market because of its high costs and the continual performance improvement of the ribbon heater.

The concept of *induction cooking*, which is great for quick responses, should be outlined here as well. Because the heating process is induced by electrical fields generated in the (pot) base, the energy flow is very efficient. Thus, a higher wattage can be exploited with far lower losses than in regular radiation heating systems, leading to shorter boil-up times. Because the heat is generated *in* the pot base, the glass ceramic is mainly "reheated" by the temperature rise in the pot. Because of maximum temperatures around 300 °C under normal cooking operation, one talks generally of "cold" cooking. Still, the realistic case of a pot cooking empty (all the liquid evaporated, leaving no cooling substance) must be taken into account. Here temperatures up to a range of 450–600 °C can arise. To keep possible risks of overheating and breakage low, the use of low-expansion glass ceramic cooktop panels such as Ceran® is necessary.

Because of the high purchase cost and the need of special pots and pans, up to now induction cooking has not been able to penetrate the market on a larger scale. Still, it is much valued for its rapid regulation and control and by and by has found its place mainly in commercial cooking.

3.2.5 Gas-Fired Systems

In countries where gas is the main heating source, which is nearly 80% worldwide, a constantly rising demand for gas cooking systems with improved design and performance can be observed. The technical and aesthetic benefits of Ceran® cooktop panels for electric cooking are easily transferable to gas cooktops. Of the manifold gas-fired systems that have been presented in the last few years, above all the Inner Burner system – a new design with a highly innovative burner concept – should be pointed out (see Fig. 3.12). This system has been developed in cooperation with a leading gas cooktop manufacturer in Asia and features ultra-low pan supports, high heating power,

Fig. 3.12. Inner Burner System with Ceran® [3.3]

and even temperature distribution on high and low heating power. It is also equipped with a newly conceived temperature sensor, which, integrated in a burner, enables the user to control the temperature during the cooking process, prevents overheating when frying with oil, keeps food warm at a given temperature and thus greatly improves safety and offers convenience. The new powerful gas burners that can be run with this inner burner system lead to the same cooking performance as an electric cooking system.

3.2.6 Thermal Stress in Ceran® Cooktop Panels

The thermal expansion of many glass ceramic materials shows a temperature-dependent behavior. As indicated in Fig. 3.13, Ceran® glass ceramic cooktop panels exhibit a parabolic-like function.

The stress distribution building up during partial heating of the Ceran® glass ceramic cooktop panels in the cooking areas and their surroundings is, therefore, a function not only of the temperature distribution in this area, but also, in a decisive way, of the curve of the thermal expansion with temperature in Fig. 3.13. In particular, quite different stress distributions will develop in the cooking areas at varying times during the warming-up and cooling-down periods, depending on the level of the maximum cooking-area temperature.

To figure out these stress distributions, we consider as a model of a cooking area a thin, plane circular disk, which is heated in a concentric circular area, as shown in Fig. 3.14. The temperature distribution is to be rotationally symmetric:

$$\vartheta = \vartheta(r); \qquad \vartheta(\varphi) = \text{const.}$$

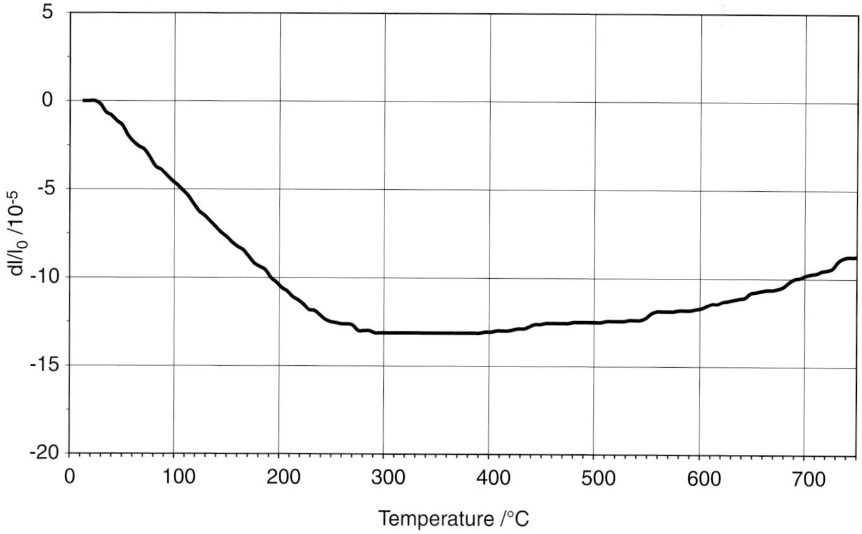

Fig. 3.13. Thermal expansion of Ceran®

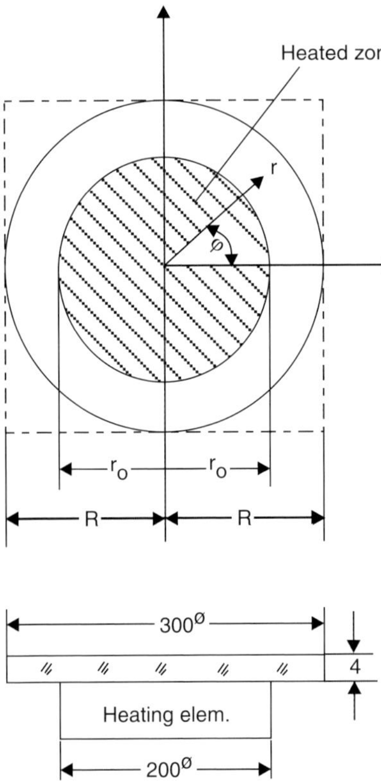

Fig. 3.14. Model of the cooking zone

For this model we further assume that the temperature will not vary over the thickness of the plate:

$$\vartheta(z) = \text{const.}$$

In consequence, in our model the stresses and displacements caused by warming-up will not vary along the thickness of the plate, resulting in a plane state of stress.

On the assumption that the thermal expansion coefficient is not dependent on temperature, the terms for the tangential and radial stresses are found in the literature [3.4]:

$$\text{tangential: } \sigma_{\text{t}} = \alpha E \left(-T(r) + \frac{1}{R^2} \int_0^R T(r) r \mathrm{d}r + \frac{1}{r^2} \int_0^r T(r) r \mathrm{d}r \right), \quad (3.1)$$

$$\text{radial: } \quad \sigma_{\text{r}} = \alpha E \left(\frac{1}{R^2} \int_0^R T(r) r \mathrm{d}r - \frac{1}{r^2} \int_0^r T(r) r \mathrm{d}r \right), \quad (3.2)$$

and, in the center of the plate, at the point $r = 0$, we obtain:

$$\sigma_0 = \sigma_\mathrm{r} = \sigma_\mathrm{t} = \alpha E \left(\frac{1}{R^2} \int_0^R T(r) r \mathrm{d}r - \frac{T_0}{2} \right) . \tag{3.3}$$

In (3.1)–(3.3), $T(r)$ is the difference between the temperatures at point r and the coldest spot of the plate.

In our case the edge of the plate is the coldest spot, and we obtain

$$T(r) = \vartheta(r) - \vartheta(R) =: \Delta\vartheta(r) \ ; \quad T(0) = \vartheta(0) - \vartheta(R) =: T_0 \ .$$

E is Young's modulus, which we assume in the following as being temperature-independent and constant.

In (3.1)–(3.3) the thermal expansion coefficient α is assumed to be independent of temperature. To consider the temperature dependence according to Fig. 3.13, we introduce thermal expansion ε by

$$\frac{\Delta l}{l} = \varepsilon(r) = \alpha \Delta\vartheta(r) = \alpha T(r). \tag{3.4}$$

From the substitution of (3.4) in (3.1) to (3.3) the terms for the tangential and radial stresses result as a function of the curve of thermal expansion:

$$\text{tangential: } \sigma_\mathrm{t} = E \left(-\varepsilon(r) + \frac{1}{R^2} \int_0^R \varepsilon(r) r \mathrm{d}r + \frac{1}{r^2} \int_0^r \varepsilon(r) r \mathrm{d}r \right) , \tag{3.5}$$

$$\text{radial: } \sigma_\mathrm{r} = E \left(\frac{1}{R^2} \int_0^R \varepsilon(r) r \mathrm{d}r - \frac{1}{r^2} \int_0^r \varepsilon(r) r \mathrm{d}r \right) , \tag{3.6}$$

$$\text{in the center: } \sigma_0 = E \left(\frac{1}{R^2} \int_0^R \varepsilon(r) r \mathrm{d}r - \frac{\varepsilon(0)}{2} \right) . \tag{3.7}$$

With the temperature distributions in Fig. 3.15 at different times of the warming-up period, and with the curve of thermal expansion of Ceran® glass ceramic cooktop panels with the temperature in Fig. 3.13, the stress distributions can be figured out. The temperature distributions in Fig. 3.15 were determined experimentally.

For doing so, we used a square cooktop panel whose lateral length coincided with the diameter of the circular disk of our cooking-area model. Along one diameter, thermocouples in the heated area were cemented into the cooktop panel. The cooktop panel was heated by a circular radiating heating element (diameter 20 cm; heating power 2100 W), and the temperatures in the various measuring points were recorded. The resulting temperatures are shown in Fig. 3.15 at the various times during the warming-up period. The temperature drop from the edge of the heating area to the edge of the plate occurs exponentially and was determined mathematically. For this, we have assumed the edge temperature to be constant at 20 °C for the period of time considered.

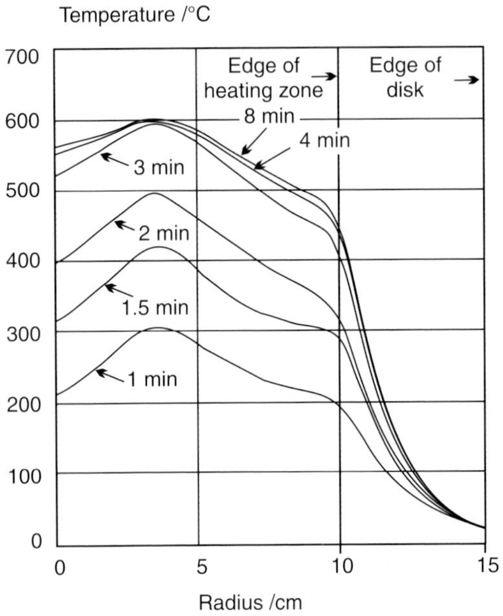

Fig. 3.15. Temperature distribution after different times of heating

We could use a square plate for the determination of the temperature distributions during the warming-up period, because the temperature distributions are rotationally symmetric also with this geometry, owing to the poor heat conduction of Ceran® glass ceramic cooktop panels (1.6 W/mK). The distortion of the temperature field by the heat dissipation into the corners of the square is negligible, regardless of the heating duration.

In analogy to the temperature distributions in Fig. 3.15 the stress distributions along a radius of our cooking-area model are represented in Figs. 3.16–3.21. The stress distributions are rotationally symmetric. The negative thermal expansion of Ceran® glass ceramic cooktop panels at temperatures below 570 °C and the subsequent positive thermal expansion (see Fig. 3.13) lead in the warming-up period (and vice versa during the cooling period) to considerable variations in the stress distribution. After a one-minute warming-up time, radial and tangential tensile stresses of about identical size appear in the hot cooking area (see Fig. 3.16). Outside the hot area, the radial stress decreases towards the edge of the plate down to zero. The tangential stress changes its sign and attains, at the edge of the plate, a compressive-stress maximum.

Half a minute later, the stress distribution in Fig. 3.17 is reached. The radial and tangential stresses start forming various stress maxima and minima. As Figs. 3.18–3.21 show, the maxima and minima become even more prominent with increasing temperatures. In the stationary state (see Fig. 3.21), the radial stresses are, to a large extent, situated within the range of the compressive stresses. Close to the temperature maximum in the heated area, the

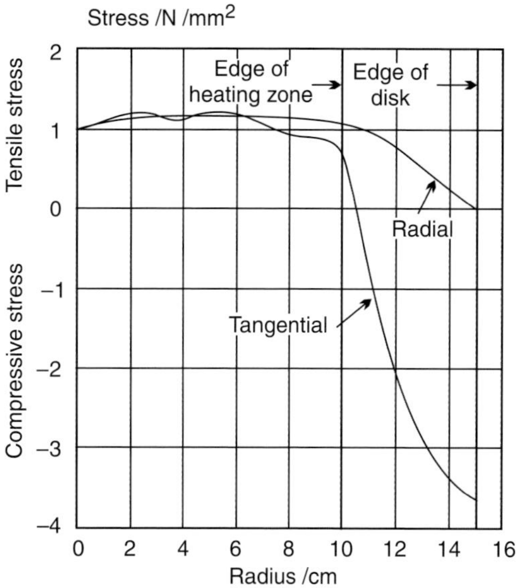

Fig. 3.16. Stress distribution after 1 min of heating

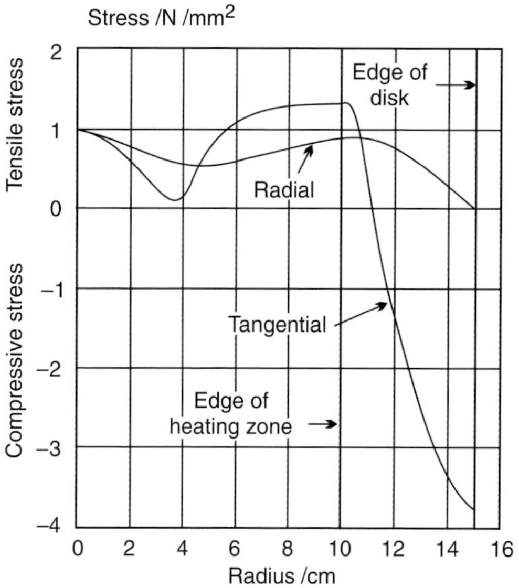

Fig. 3.17. Stress distribution after 1.5 min of heating

tangential stress reaches a prominent compressive-stress maximum, followed by a sharp maximum of the tensile stress very close to the periphery of the heated area.

As compared to glass and glass ceramics in which α is constant and positive, the following peculiarities turn up with the Ceran® glass ceramic cooktop panels:

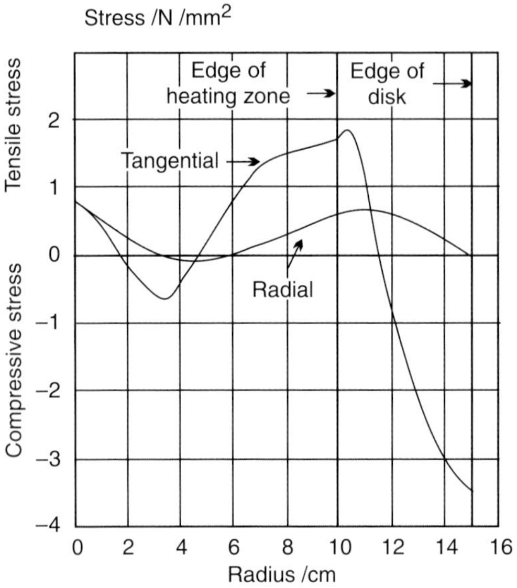

Fig. 3.18. Stress distribution after 2 min of heating

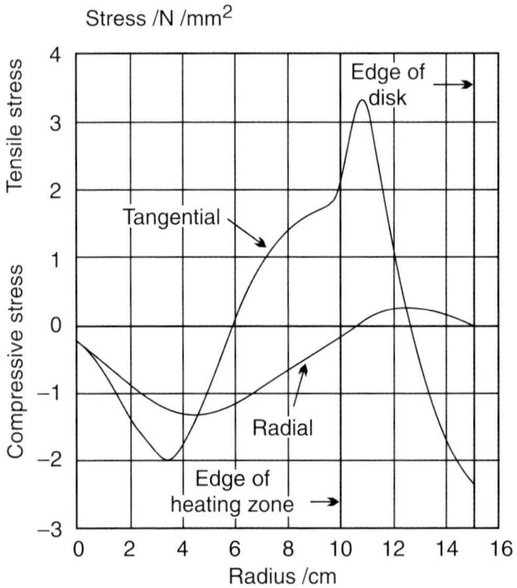

Fig. 3.19. Stress distribution after 3 min of heating

- The tensile-stress maximum of the tangential stress is located close to the border of the heated area, whereas for glass and glass ceramics with α constant and positive the tensile-stress maximum is always close to the edge of the plate.
- The radial stress likewise shows a tensile-stress maximum. In the aforementioned materials with constant and positive α, only radial compressive

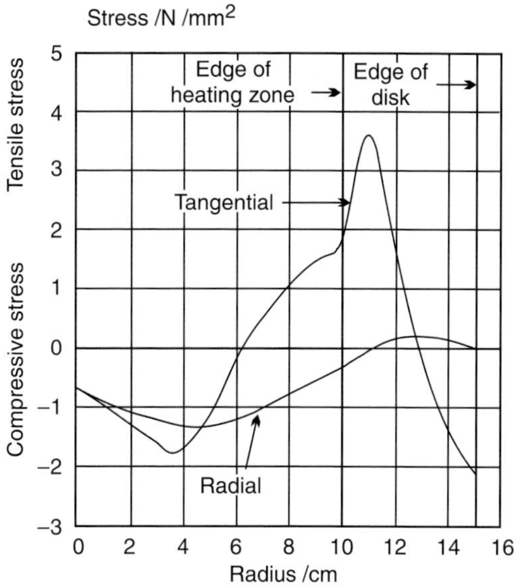

Fig. 3.20. Stress distribution after 4 min of heating

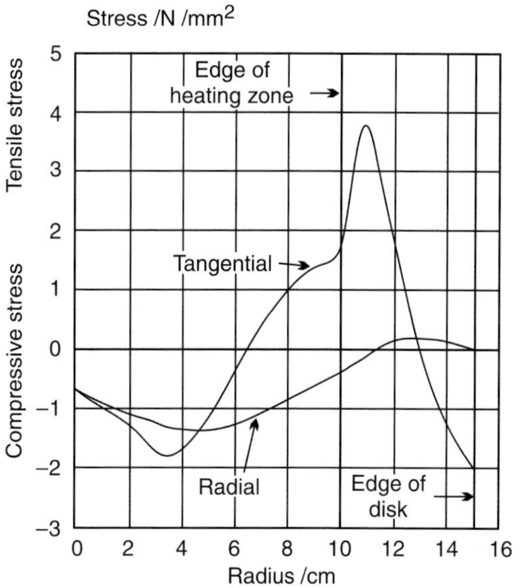

Fig. 3.21. Stress distribution after 8 min of heating

stresses appear, which towards the edge of the plate fade monotonically down to zero.

In order to review the calculation, tests were performed with circular disks and squares made from borosilicate glass and various glass ceramic materials [3.5]. As shown in Fig. 3.14, the disks were heated in a concentric circular area and heated up to fracture. The fracture mirror showed the type of stress (radial, tangential) and the distance of the maximum tensile stress from the center of the plates. The magnitude of the tensile stresses was determined by measuring the radius of the fracture mirror. With a time measurement the temperature distribution at the time of the fracture could be taken from tables drawn up before, where the temperatures averaged over the plate thickness had been inserted for the calculation. Concerning the location of the stress maximum, the correlation between experiment and calculation was very good: deviations were 2.5% for the square plate, whereas for the circular disk the values figured out were identical to the values measured. As far as the magnitude of the maximum tensile stresses was concerned, the deviation of the calculated value from the measured value was ±15% for the square plate and ±10% for the circular disk. In view of the errors possible while measuring the fracture mirror and of the definition of the temperature profile attributed, this correlation was satisfactory.

3.2.7 Compaction Stress

The structure of glass depends on the cooling procedure in the glass transition range. *Tool* [3.6] introduced the concept of the fictive temperature T_f to describe the structural state of the glass. Tool and *Narayanaswamy* [3.7] also proposed a model in order to explain structural relaxation, if the glass is heated up to temperatures T approaching T_f. In principle, this model holds for Ceran® cooktop panels, too, but contributions of the crystalline phase and the glass matrix have to be considered separately.

Structural relaxation induces volume contractions (compaction). The order of magnitude of the volume contraction depends on the thermal history of the glass ceramic as well as on the time/temperature load in practical use. The ceramization program, the cooking-area temperatures in daily use, and the holding times at these temperatures are determining factors. Because of the inhomogeneity of temperature and structural relaxation within the cooking zone and its surrounding area, the Ceran® glass ceramic cooktop panel will develop compaction stresses in course of its load time. These compaction stresses are partially reduced again by stress relaxation.

Within the heated area tensile-stress maxima of the radial and tangential stresses appear. Both are located close to the temperature maximum of the heated area. The tangential stress attains a compressive-stress maximum near the border of the cooking area. Figure 3.22 shows radial and tangential compaction stresses along the radial path from the panel edge across the edge of the heating element and towards the center of the heating element. The maximum tensile compaction stress develops with load time and depends on the load temperature as shown in Fig. 3.23.

Ceran® cooktop panels are designed to reduce the compaction stresses to a minimum. Thus, the level of stresses due to compaction is low compared to the bending strength of the Ceran® cooktops, as long as certain temperature–time load limits are not exceeded. These limits are given in Table 3.1.

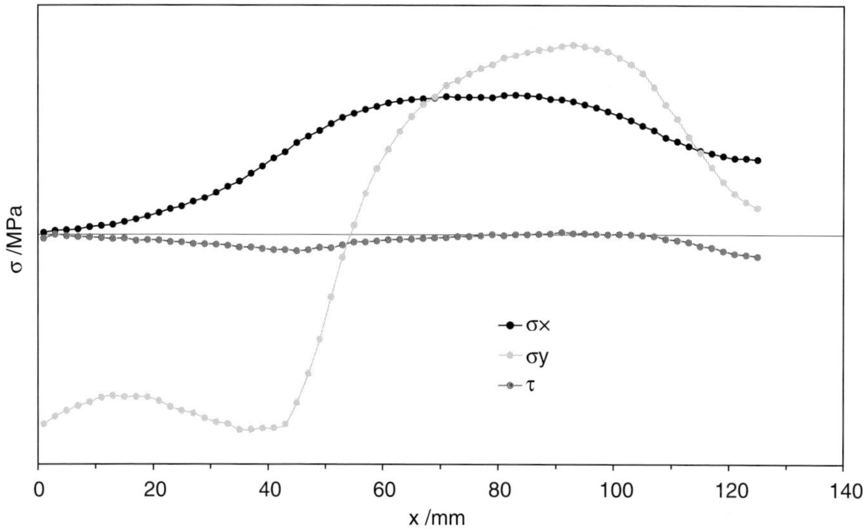

Fig. 3.22. Radial (σ_x) and tangential (σ_y) compaction stresses along the radial path x with panel edge at $x = 0$, edge of heating element at $x = 35$ mm and center of the heating element at $x = 125$ mm. Tensile stresses are positive and reach their maximum at $x = 92$ mm close to the point of maximum temperature

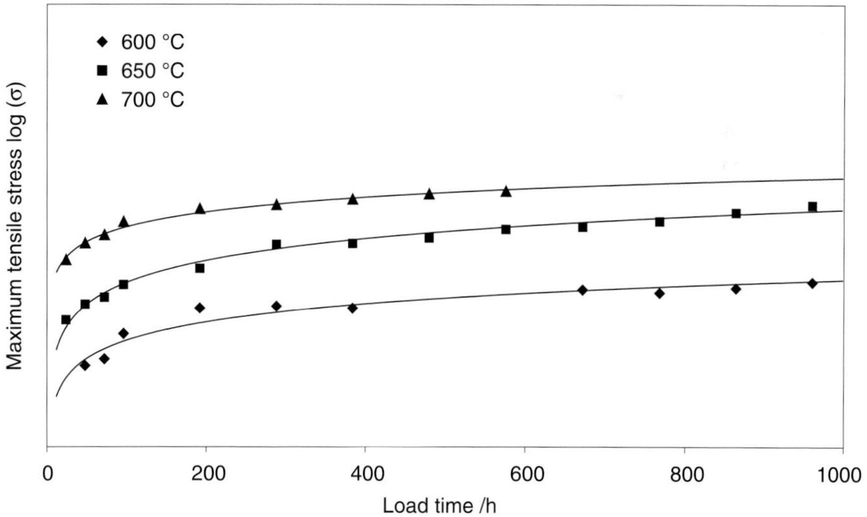

Fig. 3.23. Maximum tensile compaction stress vs. load time for different load temperatures

3.2.8 Mechanical Strength

The mechanical strength of glass is a large field of scientific investigation reaching back to the 19th century. There are numerous overviews as, for example, by *Scholze* [3.8]. Special emphasis may be given to the theory of fracture mechanics as a convincing approach to explain breakage of glass. A summary of this model is given by *Kerkhoff* [3.9].

Due to the brittleness and elasticity of glass, its mechanical strength is not an intrinsic property but greatly depends on

- the condition of the surface, i.e., kind and distribution of surface flaws,
- the time distribution of the effective tensile stress,
- the surrounding medium,
- the effective size of the surface area under tensile stress.

As a consequence, strength data given under laboratory conditions have to be transferred properly to strength data under life conditions. A concept for predicting life data is, for example, given by *Exner* [3.10].

Because there is a certain randomness to the kind and distribution of surface flaws that are typical for a certain material as well as for its production process, strength measurement data are statistical data and have to be treated in a statistical way. In most cases the strength is adequately described using the Weibull distribution.

These basic ideas of the mechanical strength of glass hold for Ceran®, too. Nevertheless, there is a variety of publications regarding those strength characteristics that refer to glass ceramics only. Most investigations of LAS glass ceramics use a fracture-mechanical approach to explain crack initiation and crack propagation in a heterogeneous system on a microscopic scale under varying load conditions (among others, [3.11–15]). Variation of chemical composition is another large field of discussion, especially in order to increase the strength [3.16, 17]. Moreover, the production process of glass ceramic itself offers extensive possibilities for strength design.

Due to the variety of effects influencing the strength of glass or glass ceramic, strength data depend on the test methods used to derive these data. For any application, predicting its resistance to mechanical loads often requires test methods close to everyday load situations. Therefore international standards as well as Schott specifications for cooktops include several tests for static and dynamic mechanical loads. Some examples for dynamic load requirements are given in the Table 3.2.

The physical behavior of glass-ceramic panels under impact varies depending on the contact situation. Accordingly, standards distinguish between flat impact (pot impact) and sharp-edged impact (steel ball impact and Norwegian hammer).

It is important to realize that the mechanical load resistance of cooktop panels is entirely determined by the strength of the bottom side. Assuming the small thicknesses of cooktop panels as usual, bending forces (caused by

Table 3.2. Standards for cooktop panels (selected)

Test according to:	Pot impact	Steel ball impact	Norwegian hammer
EN 30	x		x
EN 60335	x		x
UL 858	x	x	
UL 1026	x	x	
CSA C 22.2		x	
AS 3172	x		
SABS 153	x		x

cookware) as well as impact forces (objects falling onto the surface) bend the cooktop downwards. Tensile stresses are generated on the bottom side of the panel, whereas the top side of the panel is being compressed. Accordingly, surface flaws on the bottom side have to be avoided to a certain extent; surface flaws on the top side, such as scratches due to handling of pots or abrasive cleaning, are being compressed and therefore negligible with respect to the probability of breakage.

3.2.9 Chemical Stability

Boiled-over and burnt-on foods, as well as the cleaning agents used, are chemical substances to which the decorated cooktops – the composite of bulk material and decoration colors – must also be resistant. The applicable stipulations of the German law governing foodstuffs and consumer goods (LMBG) must also be complied with [3.18]. The decorated cooktops withstand all of these different load types without any notable changes throughout their entire useful life as long as the appliance is used according to instructions. The lifespan of such appliances now averages ten to fifteen years.

Decorated Ceran® glass ceramic cooktop panels meet the following standards:

Bulk material

- acid resistance (DIN 12116) class S2
- alkali resistance (ISO 695) class A1
- hydrolytic resistance (ISO 719) class 1

Decorated surface

- boiling citric acid (following DIN ISO 2742) $\leq 30\,g/m^2$
- hot Na-hydroxide (following DIN ISO 2745) $\leq 30\,g/m^2$

3.2.10 Electric Demands

The specific electric resistance of known glass ceramics such as the bulk material of Ceran® cooktop panels decreases with increasing temperature (see Fig. 3.24). This phenomenon is also known for glasses.

Still, cooking appliances equipped with Ceran® cooktop panels meet all known safety standards and regulations such as EN 60335-2-6, section 16 (maximum leakage current: 0.25 mA, no electrical breakdown at 3000 V) or UL 858 (maximum leakage current: 0.75 mA, no electrical breakdown at 1000 V).

In order to meet all safety standards, glass ceramic cooktops have to be built into the appliance according to specific instructions. A minimum distance of 10 mm between the heating wires and the glass ceramic panel must be kept to avoid current leakage and breakdown.

3.2.11 Summary of Physical and Chemical Properties of Ceran®

The physical and chemical properties of Ceran Suprema® are listed in Table 3.3.

3.2.12 Electronic Touches and Controls for Glass Ceramic Cooktops

An increasing number of household appliances use electronic controls (see Fig. 3.25). Cooktops especially take advantage of touch inputs. They preserve the smooth look and the sealed easy-to-clean surface. Glass ceramic is suitable because of its electrical and optical properties.

Field-effect touch technology uses the glass ceramic as an insulating substrate to propagate the electric field. The high dielectric constant of $\varepsilon_r = 8$

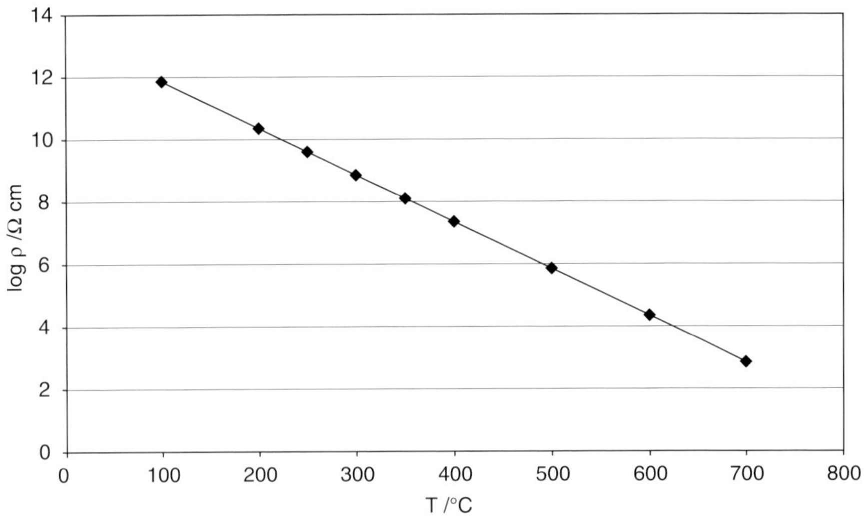

Fig. 3.24. Specific electrical resistance of Ceran® in the temperature range 20–700°C

Table 3.3. Physical and chemical properties of Ceran Suprema®

Property	Value	Dimension
Coefficient of thermal expansion	$0 \pm 0.15 \times 10^{-6}$	K^{-1}
Resistance to temperature gradients	$T_{max} \leq 700$	°C
Resistance to thermal shock	$T_{max} \leq 700$	°C
Mean specific heat capacity	0.8	J/g K
Thermal conductivity at 100 °C	1.7	W/m K
Density	2.5	g/cm^3
Young's modulus	≤ 95	GPa
Poisson's ratio	≤ 0.25	
Bending strength, bottom side	> 110	MPa
Specific thermal stress $\varphi = E\alpha/(1-\mu)$	< 0.02	MPa/K
Knoop hardness (ISO 9385)	≈ 600	
Specific volume resistivity at 300 °C	$\geq 2.0 \times 10^5$	Ωcm
Dielectrical constant (1 MHz, 25 °C)	7.5	
Loss factor $\tan\delta$ (1 MHz, 25 °C)	≈ 0.02	
Transmission, 4 mm at 600 nm	1.5–3.5%	
Transmission, 4 mm at 700 nm	19–26%	
Transmission, 4 mm at 1600 nm	$\geq 78\%$	
Color in reflection	black	
Color in transmission	orange-brown	
Hydrolytic resistance (ISO 719)	class 1	
Acid resistance (DIN 12116)	class S2	
Alkali resistance (ISO 695)	class A1	

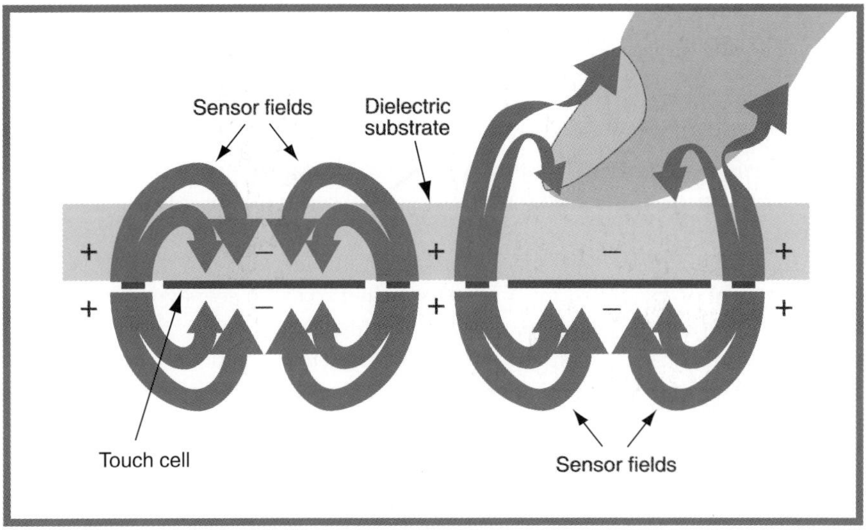

Fig. 3.25. Sensor fields of electronic touches

allows for a high penetration depth of the field. In an advanced implementation two coaxial electrodes driven by RF circuitry are used to generate and form the electrodynamic field.

The electrode design and operating frequency are chosen such that humidity from cooking and cleansers does not influence the proper function of the touch input. The field change by the electrically conductive human finger is evaluated by electronic circuitry and fed into a controller which switches the heating elements and also controls the displays.

Due to the optical transmission of the cooktop panel material in the red, readily available red 7-segment displays, red, orange and yellow single LEDs, as well as neon lamps can be used as indicators. The touch electronics is built onto a standard glass fiber-reinforced printed circuit board. This board is mounted to the bottom glass ceramic surface using very high bond adhesive tape which evens out the structure of the glass ceramic as well as the different thermal expansion coefficients of cooktop panel and printed circuit board.

The touch input is connected to a power switch ("controller") which controls the power setting of the heating elements. Thermal cut-outs on each heating element make sure that the glass ceramic temperature is kept in the intended limits. Temperature sensors can be installed to provide automatic cooking functions. A pan detection can control multi-circuit heating elements according to the size of the pan used.

3.2.13 Assembly of Cooking Systems Equipped with Ceran®

The assembly of cooking systems equipped with Ceran® is a rather demanding task. Various design guidelines have to be observed. The purpose and function of such an assembly is to create a durable joint between a low coefficient of thermal expansion (CTE) material, for example the glass ceramic used for Ceran®, and adjoining high CTE materials for housing, framing, electric components such as heaters and controls, gas ports, etc. Many designs require this joint to form a durable seal as well as a structural joint. Proper design of an elastic seal and joint requires a calculated barrier thickness correlated to CTE differentials between components and temperature gradients.

A moisture barrier between the cooking surface and the heating source, be it electrical or gas-fired, must be established to avoid seepage of liquid into the structure below. This is realized by choosing and fixing a seal between the cooktop and the housing/framing that must not deteriorate or react over time and changing environment conditions. For example, galvanized steel will oxidize over time and exposure to some adhesive systems, causing the seal and structural bond to break.

In order to meet all safety standards, the design must be resistant to breakage caused by impact, shipping, and daily use. Of utmost importance is proper handling of the Ceran® cooktop panel during assembly. Glass-to-metal contact (caused by operator contact, misuse during construction, or incorrect design) must be absolutely avoided. Pertaining to the design, the heater spring force must not be too high, nor should the structure underneath the cooktop panel force the glass ceramic to flex up.

This means the heating elements have to be appropriately arranged, fixed, and loaded (see Fig. 3.26). The complete system has to be designed and carried out in such a way that dirt collection spots are avoided. Smooth transitions between the glass and the components exposed to everyday use are prerequisites for an easy-to-clean appliance. Suitable standard sealing materials are silicone elastomers, such as room-temperature polymerizing (German: raumtemperaturvernetzende or RTV) materials, one- and two-component sealants, heat-cured silicones, and so on. Recently developed new coatings and foams are now applied to realize new appliance designs incorporating either a frameless solution or hidden seals and joints.

3.3 Development and Production of Glass Ceramic Cooktop Panels

Helga Götz, Ioannis Kosmas, Peter Naß, Erich W. Rodek, Hinnerk Schildt, Wolfgang Schmidbauer, Fritz Schröder, Fritz Siebers, Martin Taplan, Waldemar Weinberg, Evelin Weiss, Dietmar Wennemann

3.3.1 Development of Glass Ceramic Materials

Low-expansion glass ceramics are widely used as precision parts (see Chap. 4), cooktop panels, stove windows, fire-resistant glazings, cookware, and recently as reflectors in digital projection (beamers). In this section we will focus on the development of transparent glass ceramics used as fireplace windows (colorless) and cooktop panels (tinted). Newly developed translucent glass ceram-

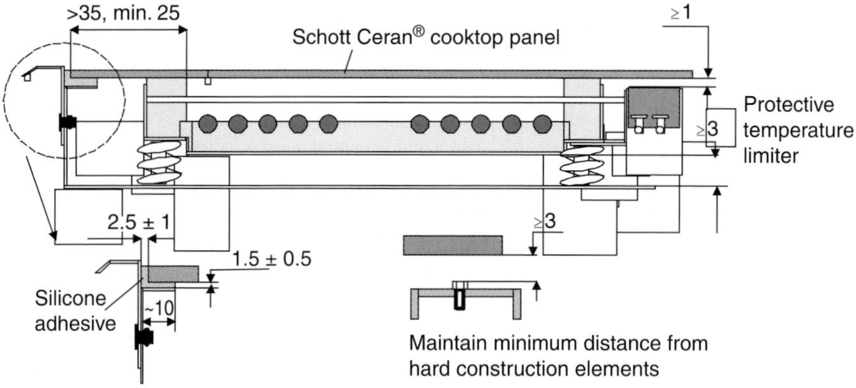

Fig. 3.26. Installation conditions for Ceran® cooktop panels

ics are also used for cooktop panels and opaque glass ceramics for cookware; both will be briefly discussed in this section.

Glass ceramics are polycrystalline solids formed from suitable parent glasses by controlled heat treatment, so-called ceramization. The glass ceramics consist of varying crystalline phases and a residual glass phase, depending on the base glass composition.

In the 1950s, *S.D. Stookey* [3.19] found that the devitrification of the parent glasses can be thoroughly controlled by adding TiO_2 as a nucleating agent. This was the breakthrough in producing glass ceramics with reproducible properties on an industrial scale and launched the development of various glass ceramic compositions for many fields of applications, as described in various review books and articles [3.20–23].

The manufacturing process of glass ceramics (for details see Fig. 3.31 and Sect. 3.3.5ff.) starts with melting and forming the parent glass by conventional techniques: the batch, consisting of natural and/or artificial materials, is melted in a glass-melting tank and afterwards the parent glass can be formed by conventional means such as pressing, blowing, rolling, or tube drawing in the desired shape. After annealing and inspection the parent glass is transformed to a glass ceramic by a two-step heat treatment, so-called ceramization (for details see also Chap. 2, Sect. 2.2). In a first step, the nucleation process, the parent glass is treated at about 20–100 K above the transformation temperature to form a nucleation phase, which acts as a substrate for the actual crystalline phase. Different oxides have been used as nucleating agents in low-expansion glass ceramics, which influence the kinetics of nucleation and the properties of the glass ceramics [3.24–29]. Nucleation is followed by a second thermal treatment, the crystallization. This process is normally performed at higher temperatures than nucleation (see Fig. 3.31). With variation of the ceramization conditions (temperature, duration) one can form different crystalline phases with adjustable sizes of the crystallites. Thus, different glass ceramics can be formed by varying the crystallization parameters.

Low-expansion glass ceramics based on the Li_2O-Al_2O_3-SiO_2 (LAS) composition have so far been the most commercially exploited system. The main crystalline phases in these glass ceramics consist of high-quartz solid solution (h-quartz s.s.), which yields coefficients of thermal expansion (CTE) ranging from about -2 to $+0.5 \times 10^{-6}\,K^{-1}$ in a temperature interval from 20 to 700 °C, and keatite solid solutions, which exhibit a CTE around $1 \times 10^{-6}\,K^{-1}$. Glass ceramics containing h-quartz s.s. are in most cases transparent and, due to the very low CTE, can be applied as cooktop panels and stove windows. The content of h-quartz (respectively keatite) solid solution crystals is controlled by varying the crystallization parameters in the manufacturing process. The crystallization of the h-quartz solid solution crystals usually takes place at temperatures between 700 and 900 °C. By further increasing the temperature or time of the crystallization treatment, the h-quartz solid solution crystals transform to keatite solid solution crystals. During this transformation the crystal size of the glass ceramic also increases from values around 20–70 nm (h-quartz s.s.) to sizes around 200 nm to 1 μm (keatite s.s.). As a consequence light scattering occurs and the visual appearance changes from transparent to translucent and finally to opaque. So depending on the heat treatment and the resulting crystal size, the appearance of keatite-s.s. containing glass ceramic can be adjusted from translucent to opaque. Transparent glass ceramics of this kind have been recently developed under the name Ceran Arcticfire™ for use as cooktop panels [3.30]. By doping the glass ceramic with coloring agents it is possible to change the appearance from white translucent for instance to a cream translucent color tone. This was done in connection with the development of Schott Ceran® Cream. Translucent glass ceramics based on keatite s.s. crystals offer new possibilities for design and decoration of cooktop panels. In their opaque form, keatite-s.s. glass ceramics have been used as cookware for many years. Some selected properties of commercially available glass ceramics of both types are shown in Table 3.4.

Depending on the application of low-expansion glass ceramics – we are now focusing on tinted cooktop panels and colorless stove windows – further physical and chemical properties are required from both the production and the user side. To lower production costs, fast and effective processing is required; for example, tank melting, on-line hot forming of panels, as well as fast nucleation rates and crystal growing are absolutely necessary. Moreover, the application of decoration, which, for example, indicates the hot zones of a final cooktop panel, should be performed at the same time as ceramization takes place, without any influence on the production quality.

The overall properties of a glass ceramic are determined by its microstructure, i.e., the percentages of the crystalline and glassy phases, the composition of the glassy phase and the type, size, shape, and composition of the crystals.

Table 3.4. Selected properties of Li_2O-Al_2O_3-SiO_2 glass ceramics (s.s. = solid solution)

Property	Field of application			
	cooktop panel	cooktop panel	stove window	cookware
CTE for 20–700 °C ($10^{-6}\,K^{-1}$)	0 ± 0.15	0.8–1.3	0 ± 0.30	1.0–2.0
Heat conduction at 100 °C (W/mK)	2.2	2.1	2.0	2.0
Max. temperature – uniform temperature distribution (°C)	700	700	700	850
Strength according DIN 52292 (MPa)	110	120	75	140
Crystalline phase	h-quartz s.s.	keatite s.s.	h-quartz s.s.	keatite s.s.

Therefore, these properties can be affected by changing the composition of the parent glass, the nucleant system, and the ceramization conditions. However, it is unlikely that we can develop a glass ceramic with several properties all near the limits of the achievable values. Thus, the properties of glass ceramics are in general the best possible compromise for the addressed application.

3.3.2 Composition of Low-Expansion Glass Ceramics

High-quartz solid solutions are "stuffed derivates" of the quartz modification of SiO_2 [3.31]. Because the SiO_2 lattice allows for many substitutions, high-quartz solid solutions in glasses can have rather complicated compositions, which may vary with the ceramization conditions. According to *Petzoldt* the composition can be described in general by [3.32]:

$$\mathrm{Li}_{2-2(v+w)}\mathrm{Mg}_v\mathrm{Zn}_w \cdot \mathrm{O} \cdot \mathrm{Al_2O_3} \cdot x\mathrm{AlPO_4} \cdot (y-2x)\mathrm{SiO_2}\ . \tag{3.8}$$

By variation of the composition of the h-quartz s.s. the CTE can be adjusted to a low absolute value for a particular temperature range, for example, less than $|\alpha_{20,700}| \leq 0.15 \times 10^{-6}\,\mathrm{K}^{-1}$ between 20 and 700 °C for use as cooktop panels.

The composition of the base glass determines the crystalline phases formed during ceramization, and influences the melting and forming behavior, the ceramization process, and the optical properties of the glass ceramic. Moreover, the base glass must meet two contradictory requirements: (a) during the melting and forming process the base glass must be sufficiently stable against devitrification, and (b) during the ceramization, high nucleation density and homogeneous crystallization have to be achieved.

Thus, commercial low-expansion glass ceramics contain at least about eight oxides, each one having a particular function. Table 3.5 shows the typical composition range and the major role of the constituents of low-expansion glass ceramics.

In order to assure the formation of high-quartz s.s. and to obtain a CTE below about $0.15 \times 10^{-6}\,\mathrm{K}^{-1}$ the LiO_2 content should be kept above about 3 wt%. This level of lithia also lowers the glass viscosity and improves the glass workability. High lithia content leads to unstable glasses and to an undesired tendency toward devitrification. The upper limit depends on the actual glass composition but is in general smaller than about 6%.

Alumina has to be kept above about 18 wt% in order to achieve a low CTE and a transparent glass ceramic. In order to avoid the detrimental formation of mullite during glass forming, the alumina content should not exceed about 25%.

The SiO_2 level is more or less constrained by that of the other components. A high silica level leads to an undesired highly viscous and seedy glass, whereas a lower content, coupled with a high alumina level, enhances the probability of mullite formation.

Table 3.5. Typical composition range in wt% and major role of the constituents of low-expansion glass ceramics

Constituent	Range wt%	Major function, influence
Li_2O	3–6	forming h-quartz s.s., CTE
Al_2O_3	18–25	forming h-quartz s.s., CTE
SiO_2	60–75	forming h-quartz s.s., CTE
MgO	0–2	forming h-quartz s.s., CTE
ZnO	0–2	forming h-quartz s.s., CTE
P_2O_5	0–5	forming h-quartz s.s., CTE
TiO_2	1–6	nucleating agent
ZrO_2	0–4	nucleating agent
Na_2O	0–2	improving glass melting
K_2O	0–2	improving glass melting
CaO	0–2	improving glass melting
SrO	0–2	improving glass melting
BaO	0–3	improving glass melting
SnO_2	0–2	fining and nucleating agent
As_2O_3	0–2	fining agent
Sb_2O_3	0–2	fining agent
Transition metals	0–2 (each)	coloring agent
Rare-earth elements	0–2 (each)	coloring agent

h-quartz s.s. = high-quartz solid solution; CTE = coefficient of thermal expansion

Because Li_2O, ZnO, MgO, and P_2O_5 are also constituents of the crystalline phase, the relative concentration of these oxides is critical for controlling the CTE of the glass ceramic. Moreover, these oxides also improve the glass working characteristics.

The substitution of ZnO by Li_2O results in a clockwise rotation of the CTE curve around the temperature reference point to lower CTEs (see Fig. 3.27, left), whereas an anti-clockwise rotation to higher CTEs is observed when ZnO is substituted for MgO. By increasing the P_2O_5 content at the expense of Al_2O_3 the expansion behavior of the glass ceramic is also significantly affected (see Fig. 3.27, right).

The most efficient nucleating agents in this type of glass ceramic are ZrO_2 and TiO_2, which are mostly added as a mixture. The total content of these agents is of the order of 3–5 wt% in order to get an efficient and fast nucleating process, which results in a high nucleation density and a transparent glass ceramic. For colorless glass ceramics, on the other hand, the TiO_2 content should not exceed about 3 wt% in order to avoid an undesirable brownish tint to the product. SnO_2 also promotes the nucleation in LAS glass ceramics.

Na_2O, K_2O, CaO, SrO, and BaO can be added to the batch in order to improve the melting behavior of the glass. However, one has to take into account that these constituents increase the amount of the residual glassy phase in the final glass ceramic, which raises the CTE and may cause undesirable haze in the transparent material.

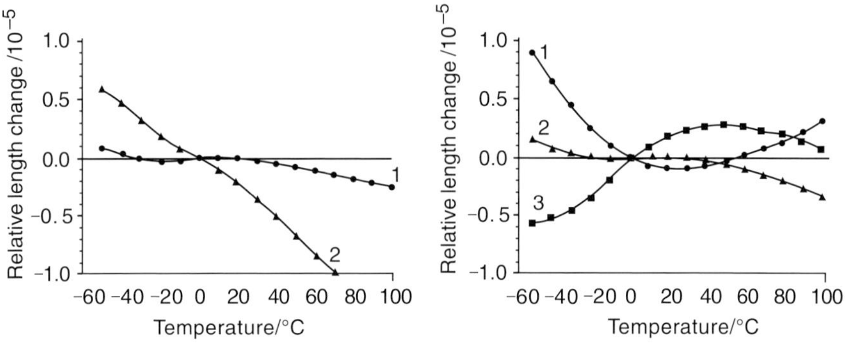

Fig. 3.27. *Left*: rotation of the $\Delta l/l$ curve ($1 \longrightarrow 2$) by substitution of 0.1 wt% of ZnO by Li_2O. *Right*: changes in the $\Delta l/l$ curve by increasing the P_2O_5 content at the expense of Al_2O_3 for curves 1–3 [3.33]

Finally, As_2O_3, Sb_2O_3, or SnO_2 are added to the batch as fining agents in order to reduce seeds in the glass article.

As mentioned before, glass ceramics used as cooktop panels have, furthermore, to meet specific transmission specifications in the visible (nearly opaque) and in the near-IR (transparent) spectral regions. Different well-known colorants, such as Fe_2O_3, NiO, CoO, Cr_2O_3, CuO, MnO_2, V_2O_5, and CeO_2 can be added to the glass composition in order to achieve the required transmission characteristics. The doping level is typically less than about 1 wt%. The coloration of LAS glass ceramic is described in detail in the patent literature (e.g., [3.34–38]). In Fig. 3.28 the transmission of a CoO- and MnO_2-doped sample is shown in the visible spectral region before and after ceramization. It is obvious that during the ceramization the transmission spectra and, therefore, the color is changing. The manganese-doped glass changes its color from colorless (flint) to yellow and the CoO-doped glass from blue to lavender. These color changes, which are also observed with other dopants, are still not completely understood, but can probably be attributed to enhanced scattering from the formed crystallites, to changes of the valence states, and/or to changes of the actual sites of the coloring ions.

Ceran® Color, formerly produced at Schott AG, was colored by a mixture of Fe_2O_3, NiO, CoO, and MnO_2. Ceran Hightrans® is colored with V_2O_5, does not have high Fe_2O_3 levels, and exhibits, therefore, a substantially improved infrared transmission.

The higher infrared transmission is a key factor to improve the cooking performance because the boil-up time is reduced. Higher infrared transmission allows the heating energy in the form of radiation to pass through the cooktop panel to the cookware more efficiently. Striving for even further optimization of the cooking performance a second generation of high-IR transmitting glass ceramics was developed. This new generation of glass ce-

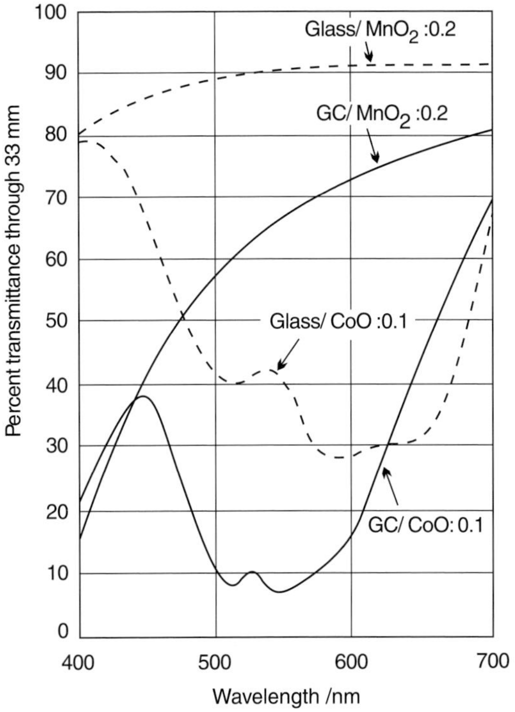

Fig. 3.28. Transmission of a parent glass (dashed line) and a glass ceramic made thereof (solid line), doped with CoO and MnO_2 [3.34]

ramic cooktop panels was recently introduced into the market under the name Ceran Suprema® (see also Sect. 3.2.3 "Transmission and T–t Loading").

3.3.3 Decoration

Surface decoration is needed for several reasons: the smooth, shiny surface is rendered even more attractive by suitable decoration. Cooking appliance manufacturers can raise their product profile with attractive decoration of this kind and achieve a valuable product differentiation providing their own design line. The practical aspects are perhaps most important: the decoration clearly marks the position of the cooking zones even when the heating elements are switched off. Decoration makes surface scratches, fingerprints, etc., less noticeable. The thickness of the decoration material prevents or considerably reduces the risk of damage to the surface caused by cookware with rough bottoms, for example.

Suitable ceramic colors have been developed that stand up to the considerable everyday wear to which decorated ceramic cooktops are subjected. In most cases, high temperatures, aggressive media (boiled-over food, any residues of chemical cleaning agents), and mechanical stress (rough cookware, abrasive cleaning agents or additives) all act upon the surface at once. This means that the decoration material itself, as a part of the surface, must resist without substantial damage/deteriorations all the loads for many years. Among modern products, only ceramic colors developed specifically for this purpose can fulfill this demand.

3.3.4 Composition of Ceramic Colors

As mentioned above, Ceran® glass ceramic cooktops are decorated with ceramic colors of varying chemical composition to achieve an extensive color selection.

Ceramic color consists of a vitreous matrix ("flux") in which inorganic bits of color ("pigments") are embedded. The chemical composition of the flux may show a very complex chemical composition, but it is usually a transparent, colorless glass. The pigments dispersed in the flux produce the actual color effect. Only pigment types that are not attacked (dissolved) by flux can be used. For this reason, feasible flux/pigment combinations are limited in number. The pigments used in the fluxes mentioned above are, for the most part, of oxidic composition, often with a spinel structure ($A^{2+}B_2^{3+}O_4$, with, e.g., a cubic crystalline structure). The coloring effect of these so-called "dispersion-clouded ceramic colors" is based on light refraction and dispersion as well as absorption by the pigments.

The sum of the load exposure factors involved, which hardly ever occur alone, indicates that the "glass ceramic/ceramic colors" material composite as used on cooktops is among the materials with the highest load exposure levels in the modern world. Achievement of the present quality level was only possible on the basis of sophisticated knowledge and experience, extensive goal-oriented development of chemical combinations optimizing the interactions between color flux and glass ceramics, and application of Schott's own practical testing methods.

The following describes in detail the stresses to which a decorated Ceran® cooktop is exposed. The ceramic colors usually used on glass ceramics have a thermal expansion coefficient (α) of approximately 3–12×10^{-6}/K (Schott measurements). The glass ceramic of Ceran® cooktops itself has an $\alpha(20/700)$ of $0 \pm 0.15 \times 10^{-6}$/K. When the colors are fired, the resulting combination of the two materials leads to considerable tension in the contact zone, causing chipping-off of the colors along with a thin layer of glass ceramic following firing. The relevant tensions are calculated according to (3.9), (3.10), and (3.11) as follows [3.39]:

$$\epsilon_{\mathrm{GK}} = \frac{(\alpha_{\mathrm{GK}} - \alpha_{\mathrm{F}})(T - T_0)}{1 + [(1 - \mu_{\mathrm{F}})/(1 - \mu_{\mathrm{GK}})][(E_{\mathrm{GK}}\, d_{\mathrm{GK}})/(E_{\mathrm{F}}\, d_{\mathrm{F}})]}\,, \tag{3.9}$$

$$\epsilon_{\mathrm{F}} = \frac{(\alpha_{\mathrm{F}} - \alpha_{\mathrm{GK}})(T - T_0)}{1 + [(1 - \mu_{\mathrm{GK}})/(1 - \mu_{\mathrm{F}})][(E_{\mathrm{F}}\, d_{\mathrm{F}})/(E_{\mathrm{GK}}\, d_{\mathrm{GK}})]}\,, \tag{3.10}$$

resulting in the following tension stresses:

$$\sigma_{\mathrm{GK}} = \frac{\epsilon_{\mathrm{GK}}\, E_{\mathrm{GK}}}{1-\mu_{\mathrm{GK}}} ,$$

$$\sigma_{\mathrm{F}} = \frac{\epsilon_{\mathrm{F}}\, E_{\mathrm{F}}}{1-\mu_{\mathrm{F}}} . \tag{3.11}$$

Here ϵ is the expansion ($\Delta l/l$), σ is the tensile stress, $-\sigma$ is the compressive stress, α is the average linear coefficient of thermal expansion, E is the modulus of elasticity, d is the thickness, μ is Poisson's ratio, and $T - T_0$ is the temperature difference hot/cold. The indices are for glass ceramic (GK) and for color (F) respectively.

The calculations show that only a very thin color layer (< 5–7 µm) can be applied so as to adhere, since otherwise the tension threshold is exceeded, leading to chipping-off of the composite. This makes it necessary to produce colors that are opaque at such minimum layer thicknesses, which can lead to difficulties in view of pigment particle sizes often exceeding 10 µm. Another difficulty is that the upper cooktop surface of the colors deemed generally suitable for decoration of Ceran® cooktop panels must also withstand other – sometimes extreme – thermal, mechanical, and chemical loads (see also Sect. 3.2).

The colors themselves vary widely and are basically pigmented multi-component systems containing up to 15 components. According to the main components in the glass flux these modern decoration colors are referred to as *a*lkaline-*b*oron-*s*ilicate formulations. Other components (Al_2O_3, earth alkalines, ZnO, ZrO_2, TiO_2, etc.) are added. These complex formulations are necessary to achieve the outstanding specifications for this application without environmentally questionable components such as Pb, Cd, formerly used in other industries [3.40].

The precise chemical formulas for individual colors – the result of exhaustive developmental work involving statistical trials and constant optimization – are of course commercial secrets. Table 3.6 lists some of the physical properties of the colors.

3.3.5 Production

The manufacturing process for Ceran® cooktop panels can be divided into two fundamentally different partial processes. The first process, the production of the parent glass, is basically not different from the production process

Table 3.6. Physical properties of Ceran® decoration colors

Transformation temperature T_g (°C)	450–600
Working temperature V_A (°C)	800–1100
Coefficient of thermal expansion $\alpha_{20/300}$ ($10^{-6}\,K^{-1}$)	3–12
Densification temperature (°C)	500–700

for other technical glasses (see Fig. 3.29). The well-known process steps of molten glass, shaping, annealing, and cutting of the raw glass are followed. The second process involves the mechanical finishing of the parent glass (cutting, grinding, drilling), the decoration with special colors (see Sect. 3.3.3), and ceramization (see Fig. 3.30). Only after ceramization, during which the homogeneously distributed crystals are formed in a thermal process, are the special properties of the Ceran® cooktop panels achieved, for example, the excellent resistance to temperature shocks and practically negligible thermal expansion. A schematic drawing of the temperature curve in production is shown in Fig. 3.31.

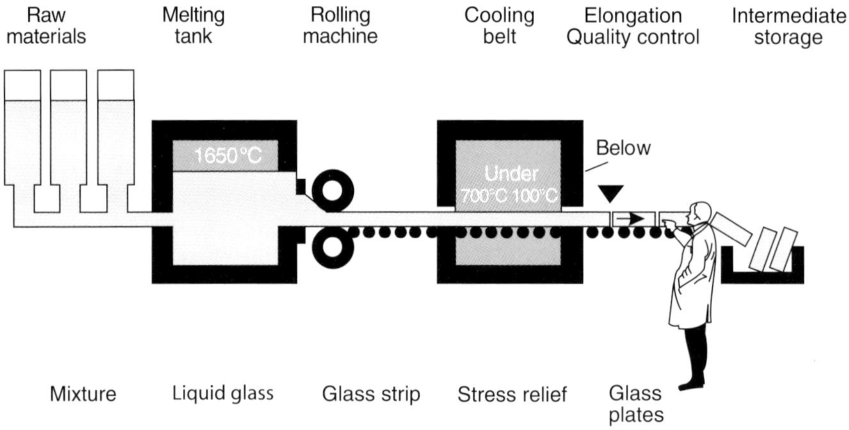

Fig. 3.29. Glass manufacturing

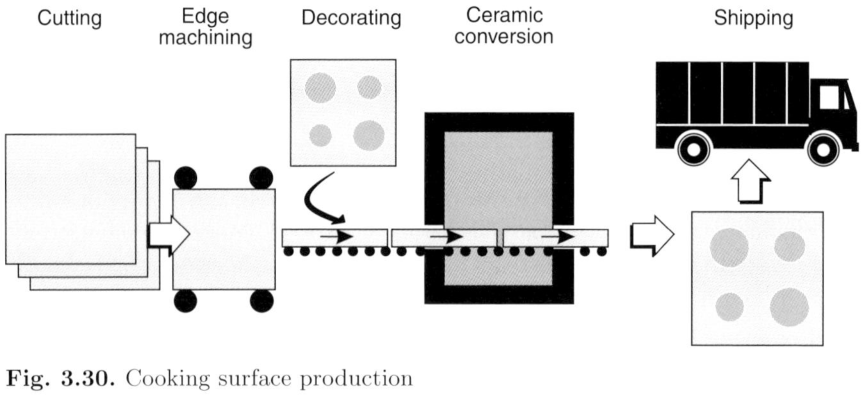

Fig. 3.30. Cooking surface production

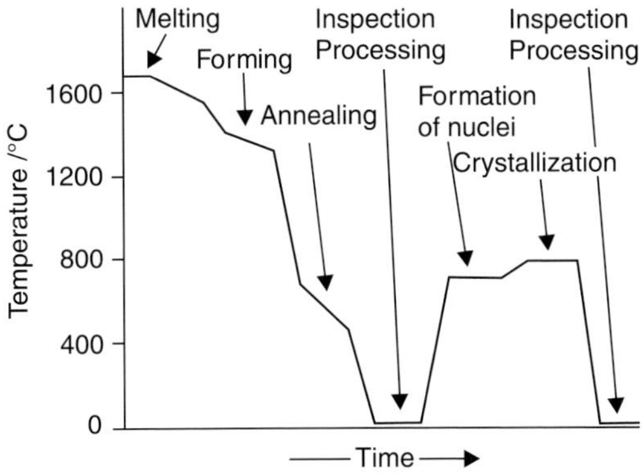

Fig. 3.31. Manufacturing schedule for glass ceramics

3.3.6 Batch Materials

The raw materials for the melt of the parent glass of Ceran® cooktop panels are selected with regard to the strict demands placed on the purity and melting properties of the available natural and synthetic compounds. In accordance with the composition, the main components, quartz, aluminum hydrate, lithium carbonate, titanium oxide, and zirconium oxide, are mixed together. Further additives are soda, potash, zinc oxide, magnesium oxide, barium nitrate, and a refining agent. The transmission of glass ceramic in the visible and infrared (IR) range is achieved by adding the coloring oxides Fe_2O_3, CoO, NiO, MnO_2, and V_2O_5 in the required amounts [3.41].

Apart from purity, a suitable grain size is also important when selecting raw materials. Individual large grains and agglomerates must be avoided to the greatest extent possible, because they can be a source of stones and, in some cases, SiO_2-rich lumps. In particular, the raw material SiO_2, which is present with more than 60% by weight and has a melting temperature of more than 1700 °C, must fulfill stringent requirements with regard to purity and fineness.

All raw materials are continuously subjected to a thorough analysis. Any fluctuations in the raw materials are taken into consideration when calculating the weighed-in quantities of the batch.

Cullets are the second component of the batch. In-house cullets, which are generated partially by unavoidable process disturbances or which accumulate during cutting and mechanical processing, are used for the parent glass melt of Ceran®. Cullets are an indispensable aid in the melt, acting as a fluxing agent and accelerating the dissolution of the sand. This allows the melt output to be increased, whereby the specific energy consumption is reduced [3.42]. Depending on the supply, cullets can account for 30–50% of the batch.

The batch house is solely responsible for the raw materials and cullets. Through cooperation with the Purchasing, the Research and Development, and the Melting Departments, a reliable supply of materials for the melting process is ensured. The computer-assisted calculation of the glass batch, taking the current raw materials analyses into consideration, and automatic batch weighing-in are indispensable aids.

3.3.7 Melting

The most important phase of glass production is the melting process. At temperatures up to 1650 °C, the crystalline raw materials and the cullet combine to form a homogeneous glass. The structure of the melting unit, the selection of the refractory material, the type of heating, and the melting temperature control are attuned to the parent glass of Ceran® and ensure the necessary melting output and quality of the glass. Many years of optimization work have led to continuous improvement of the melting units. A description of the standard melting processes can be found in [3.43].

The parent glass of Ceran® is melted in a continuously operating tank furnace. High-grade, refractory materials on a ZrO_2–SiO_2 basis are used for the tank lining [3.43]. The tank is subdivided into a melting section, refining section, and fore hearth.

The reaction of the raw materials during the glass formation process places significant stress on the tank lining refractories in the melting section. The glass formation is accompanied by the release of gaseous raw material components, mainly water, carbon dioxide, and nitrous oxide. In this area, small amounts of lightly volatile components, such as the refining agent, evaporate and very fine particles of the raw materials are carried along as dust. In order to reduce air pollution, both the nitrous oxide and the solid components in the exhaust gas need to be restricted below legally defined limiting values. The nitrous oxides are removed in a reaction with ammonia [3.44]. The solid components are removed using special high-temperature filters and are added again to the melt as weighed-in components.

The highest temperatures are attained in the refining section, where the bubbles trapped in the glass should rise to the surface and thus be removed. Because in particular the small bubbles only have a low rising force and, therefore, a slow velocity of ascent, the glass needs to remain in the refining section for a sufficiently long period of time. The refining process is supported by refining agents, which release gases in the refining section and support the formation of larger bubbles with increased velocity of ascent [3.45].

From the refining section the glass flows into the fore hearth. Here, the glass is cooled to the required processing temperature and mechanically homogenized in a stirring system. The fore hearth is also built using especially corrosion-resistant and heat-resistant refractories, which do not lead to quality losses in the glass due to striae or stones. In an ideal situation, the shaping section is provided with homogenous glass that is free of bubbles, striae, and crystalline inclusions.

Gas, heating oil, and electricity are used for heating the tank [3.43, p. 208]. Depending on the tank range, electricity or fossil fuels are preferred. The combustion air is preheated by the waste gas via heat exchangers in order to reach the required flame temperature and to reduce energy consumption. Recently oxyfuel technology was introduced and give extended potentials for further development and energy reduction.

The comprehensive use of modern data-processing technology is used to fulfill these demanding requirements of the Melting Department. Using a process management and control system, all process-relevant data are recorded and analyzed. This includes the temperature at various points in the melting tank, electric heating data, the gas, oil, and air quantities, and additional variables important to the process, for example, the melting output. These data are combined with data on the product quality in databases and can be evaluated for optimizing tank operation and used for troubleshooting by graphical and mathematical methods.

3.3.8 Hot Forming of Glass Ceramics

The shaping of the parent glass of Ceran® is a very complex process, which differs from the familiar, conventional roller process. The main differences as compared to conventional rolling processes/techniques lie in: (a) feeding the glass through a nozzle, (b) the high temperature of the glass (which introduces high thermal stress on the rollers) and, (c) the need to achieve a homogeneous glass surface.

Consistent process analysis and development made it possible to establish a stable shaping process for Ceran® cooktop panels. The emphasis of this process/technology development was on the glass nozzle in front of the processing roller, the roller material (including surface treatment), and the controlled heat dissipation between the glass and the processing rollers. The most important issue of this development was to assure that no uncontrolled crystallization of the glass takes place during the shaping process.

The parent glass melt is cooled down to the processing temperature in the fore hearth, homogenized and fed to the rolling machine via a nozzle. The glass flow is controlled by the size of the nozzle gap and the roller velocity.

The rolling machine shapes the glass flowing to the process roller into a glass strip. During shaping, controlled heat dissipation from the glass is achieved via the processing roller until the required shape stability of the glass strip has been attained.

The heat dissipation during shaping is very complex and needs to be coordinated exactly. The heat flow between the glass and the rollers should be performed such that the glass can continue to flow during shaping. Thus, the glass surface temperature may drop below the transformation temperature for only a short period of time. The glass strip that then leaves the roller becomes viscoelastic again due to reheating from the core. The heat potential in the core of the glass strip leads to renewed heating of the glass surface and, therefore, to flux of the glass for a short period of time. The final dimensioning with regard to the glass thickness and the strip flatness takes place in this stage.

The reheating is concluded quickly due to further cooling of the glass strip via heat loss to the air and transport rollers (see Fig. 3.32). The glass becomes rigid and the shaping process is completed.

The complex thermal processes during shaping in the rolling machine can be described as follows: the heat flow between the glass and the rollers is by means of conduction, convection, and radiation. The course of heat flow across the circumference of a roller is reflected in the heat-flow diagram (see Fig. 3.33).

The local temperature distribution and the change of the temperature on the surface of a roller with time can be viewed as the overlaying of a stationary temperature field with a field resulting from a temperature oscillation. The illustration shows the local heat-flow density in a roller as a function of the number of revolutions (see Fig. 3.33).

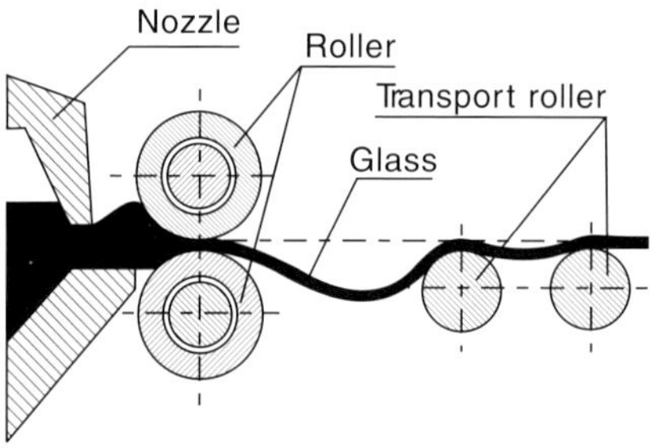

Fig. 3.32. Shaping of the glass strip

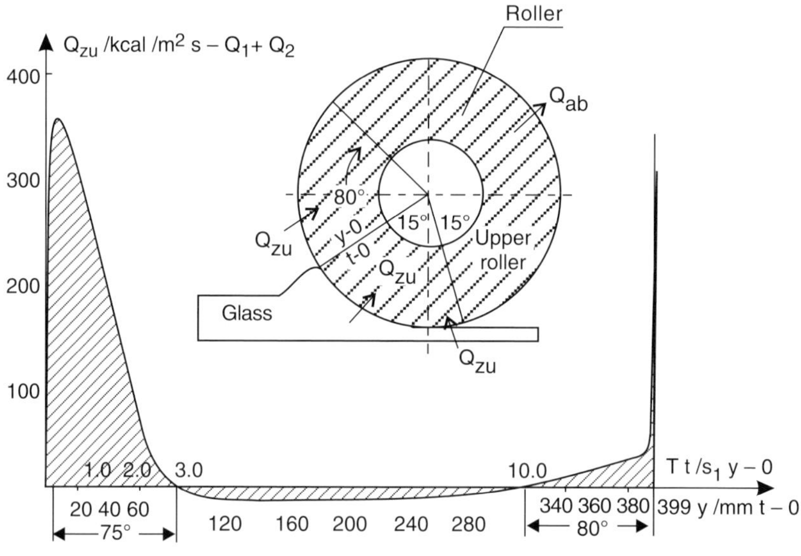

Fig. 3.33. Local heat-flow density according to [3.46]

Shortly after the glass contacts the roller, the heat-flow density peaks. The subsequent drop of the heat flux is due to the smaller temperature difference between the glass and the rollers and to the reduction of the heat conductivity of the glass with decreasing temperature. As shown in the diagram, most of the heat is transferred through conduction following the first contact between the glass and the roller, where a sector of about 75° is in contact with the glass.

3.3.9 Annealing

The further course of the process following shaping entails transporting the glass strip into an annealing furnace via a roller table. In the roller annealing furnace, the glass is kept at temperatures higher than the transformation temperature but lower than the temperature at which nucleation and crystal growth occur. The internal stresses of the glass introduced during the rolling process are released during this annealing process to values allowing for further processing of the glass panels. After annealing the glass strip in the roller furnace, it is transported to the cutting and breaking device via a roller conveyor.

3.3.10 Glass Cutting and Stacking

After the glass strip is transported out of the roller furnace, glass cutting, quality testing, and stacking of the raw glass in racks ensues. The production/machine technology required for these process stages mainly complies with the state-of-the-art technology. In individual cases, specific new developments were necessary to process the raw glass. For example, modifications of the cutting and breaking technology to meet special requirements were necessary. As the conventional processes did not lead to the required cutting and breaking quality of the glass panels, modifications of the cutting tools for scoring as well as for the breaking process for the boart break had to be developed. These special requirements for the processing of the glass strip had to be addressed during the development of the machine technology.

The machines feature modern CNC drive technology for controlling the machine axes. The cutting and breaking machines and the stacking device are controlled and monitored by a process management and control system. The setting parameters for the machines and the transport systems are set to the respective panel dimensions without long production interruptions using this flexible machine control system. The panels cut to customer specifications are machined fully automatically in the aforementioned production stages and are transferred to the raw-glass warehouse after being placed on racks. A high level of know-how in process/production technology and continuous quality control in shaping the parent glass ensure the product quality demanded by the market. Thanks to the continuous development/improvement of the production processes/technology, the product demands expected from the market in the future will be fulfilled as well.

3.3.11 Mechanical Processing

During the mechanical processing of the parent glass, machined and decorated, yet still glassy panels are made out of the raw plates.

Mostly state-of-the-art glass working machines are used for the mechanical processing, i.e., cutting, grinding, drilling, and edging. Depending on customer requirements, a wide variety of nearly endless shapes can be realized within the maximum dimensions of $980 \times 1400\,\mathrm{mm}^2$. This includes contour-ground forms as well as drillings. The requirements for high mechanical strength of the cooktop panels call for special machining tools and process parameters adapted to the special parent glass. A reduction in strength due to improper processing is unacceptable. Numerically controlled machines and statistical process control ensure the fulfillment of these strict demands.

A variety of shapes can be manufactured individually on special customer demands, such as

- holes for gas burners,
- holes for control knobs,
- various edge designs, such as beveled (bevel size up to 33 mm), c-shaped or rectangular,
- oval and round panel shapes.

Newer designs embrace

- finger depressions to more easily find the touch point and
- ground grooves as spill containment.

3.3.12 Secondary Processes

In recent years, customers increasingly have been asking for new appealing designs that could be realized by shaping or bending.

Consequent analysis and development of thermal processing of glass and glass ceramics brought up new ideas that disclosed exciting possibilities to create a "new look".

The glass ceramic wok shown in Fig. 3.34, a product that serves the rising demand for Asian style cooking, is manufactured by a method of vacuum-based hot forming in a glassy state.

Another version of wok with an incorporated flat panel, displayed as a single cooktop, is realized by applying *NIR-based hot forming*, a highly sophisticated process, where the glass plate is uniformly reheated and shaped by using infrared radiation.

The key to success of this unique method is the high energy input into glass. With NIR (near-infrared) radiators built in an appropriate radiation

Fig. 3.34. Ceran® glass ceramic wok

cavity, as pictured in Fig. 3.35, a high energy input by a specific combination of absorption and reflection is realized. This is visualized in Fig. 3.36, showing the transmission spectrum of Ceran Hightrans® in comparison to the spectral emission of two radiator types.

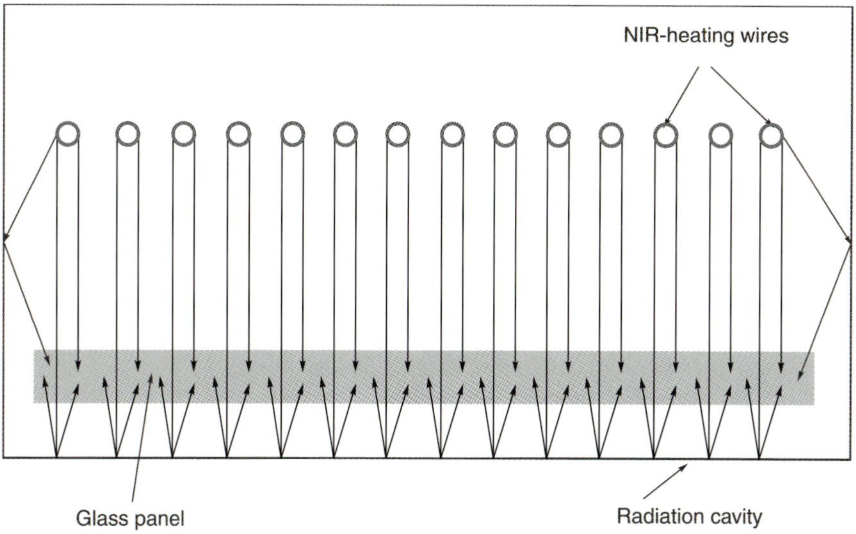

Fig. 3.35. NIR radiator in a radiation cavity

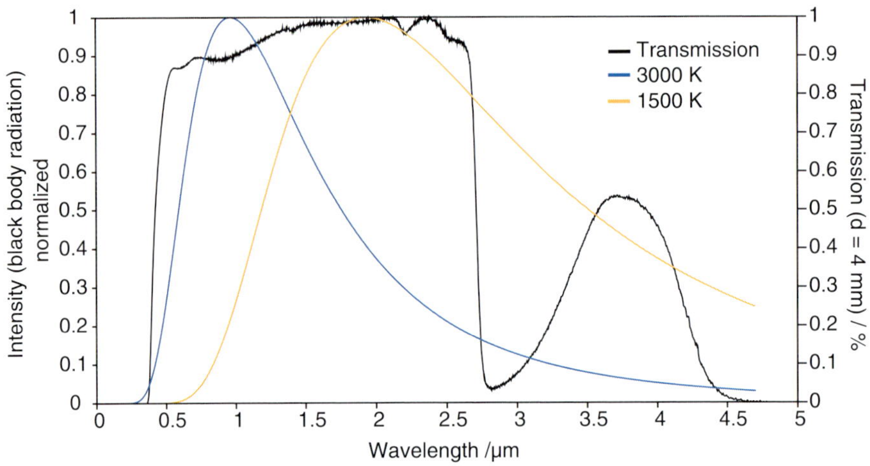

Fig. 3.36. Normalized transmission spectrum of Ceran Hightrans® in comparison to the spectral emission of two radiator types

The radiation of a 1500 K standard heater (yellow graph) will be absorbed in a thin surface layer of the glass panel, leading to an inhomogeneous temperature profile in the panel. The specific absorption of the NIR heater (blue graph, 3000 K) is lower. This combined with a multiple reflection of the radiation within the cavity walls leads to a thorough and uniform heating of the glass panel. The benefits of this method are homogenous and fast heat-up rates allowing a very rapid passing of the critical (i.e. crystal forming) temperature range.

Further prominent design features, which can be realized with this shaping method, are:

- gas ranges with bent-up rims around burner holes (like a volcano) (see Fig. 3.37a),
- *downdraft tray* as spill containment (see Fig. 3.37b),
- various innovative 3D shapes in the range of Robax® fireplace sight panels that meet the current design trend going to bigger panel sizes and round bends such as the *Dome* and the *Elisse* (for pictures see also Sect. 3.4).

A *combination of mechanical bending and thermal shaping* leads us to so-called bent shapes that help create an arrangement of cooktop with panel incorporating control and display panel into an all-in-one Ceran®-shaped feature. The laboratory hotplate series from Schott Instruments, for example the *hotplate unit SLK 6* with a bent heating surface (see Fig. 3.38) and the bent cooktop panel *Voss Futura Vision*® from Electrolux, as well as the vast variety of transparent Robax® fireplace sight panels with bending radii of up to 90°, are outstanding examples of the successful application of this method. The design features of Swingline® (Fig. 3.39) give a stylish touch

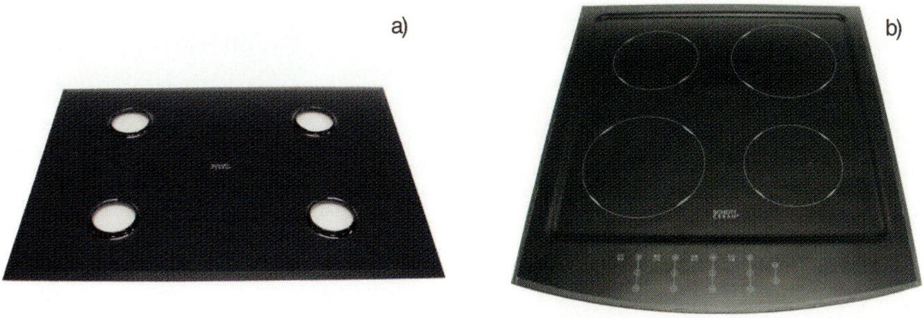

Fig. 3.37. (**a**) Gas range with bent-up rims around burner holes and (**b**) downdraft tray

Fig. 3.38. Laboratory hotplate unit SLK 6 with a bent heating surface [3.47]

to kitchen ranges. This cooktop panel is realized by using *gravity force at elevated temperatures.*

3.3.13 Screen Printing

The mechanically processed panels are embellished with special decorative colors in accordance with customer wishes. The decoration design serves both as a means of delineating the cooking zones and for realizing the special designer wishes of the customers. The application of decorative colors specially matched to the Ceran® cooktop panel is done using conventional technologies, such as screen-printing, wet application, or heat release. Sometimes, however, more extensive modifications and adaptations of the process steps are required.

Fig. 3.39. Cooktop panel Swingline® [3.48]

3.3.14 Ceramization Process

The elementary processes, which determine the properties of the glass ceramic following the primary annealing of the base glass and quality control, are nucleation and crystallization. The characteristic T–t function which defines these processes is pictured in Fig. 3.31, see Sect. 3.3.5.

As mentioned in the introduction, the ceramization of the parent glass is performed by a separate two-step heating process: nucleation and crystal growth. During nucleation, nuclei are created by phase separation of the nucleating agents. These nuclei are necessary as the starting material for the crystal growth. Because substantial nucleation takes place at lower temperatures, more so than crystal growth (see Fig. 3.40), no crystals can be formed during the cooling process of the parent glass from the melt. However, if the material is heated up in a second step, nuclei are created at a typical rate of 10^{14}–10^{16} nuclei per cm^3 per minute, and at higher temperatures crystal growth can take place. The maximal growth rate is of the order of 5–100 μm/min.

However, the actual temperature program of the ceramization process is a complex function of time and temperature rather than a simple two-step process, because one has to take into account the changes in transmission, viscosity, density, and CTE during the crystallization.

To get transparent tinted or colorless glass ceramics one has to avoid strong scattering from the crystal sites. The scattering intensity scales with

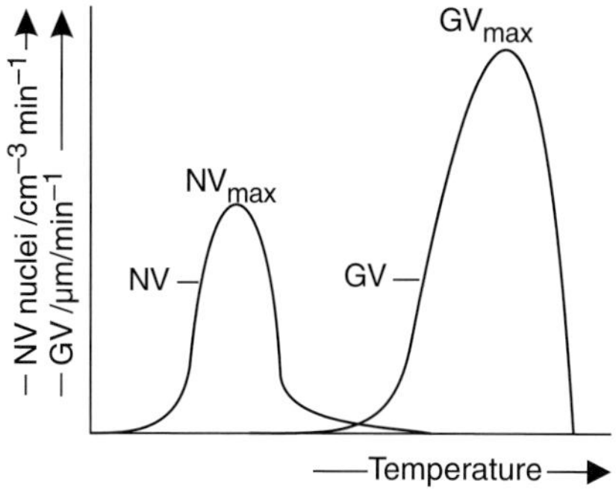

Fig. 3.40. Nucleation and crystal growth velocity of parent glasses as a function of temperature

$\eta_{\mathrm{vol}}\,(\Delta n)^2\,a^3/\lambda^4$, where η_{vol} is the volume percentage of the crystal phase, Δn is the difference of the indices of refraction of the glassy and crystalline phases, a is the crystal radius, and λ is the wavelength. Hence, to avoid substantial scattering losses, for given η_{vol} and Δn, the crystal size a should be as small as possible. It is demonstrated that a nucleation and, subsequently, a crystal density of about $10^{16}\,\mathrm{cm}^{-3}$ is sufficient to restrict the crystal size to below about 100 nm. In order to obtain these high nucleation densities one has to provide a sufficient amount of nucleating agents and to remain in the temperature region of the nucleation for a sufficiently long period.

To figure out the right temperature regions for nucleation and crystal growth one often applies the straightforward technique of annealing thin glassy rods in a gradient furnace. The rod is placed for a certain time in a gradient furnace, which covers the assumed nucleation and crystallization temperature interval. After the heat treatment, different phases of the material can be easily distinguished along the rod by variations in color and transparency. In Fig. 3.41 the method and results are shown schematically for a typical LAS glass ceramic [3.49].

At low temperatures the glassy state is still present, followed by a small transparent region with small high-quartz s.s. crystals. At higher temperatures the keatite solid solution is formed as mentioned in the introduction. Because these crystals are larger, the glass ceramic becomes opaque due to strong scattering from the crystallites. The corresponding changes in the CTE are also indicated in Fig. 3.41.

The crystallization process itself is an exothermic reaction (on the order of 0.1 J/kg) and accompanied by a strong increase in viscosity (3–4 orders of magnitude). Moreover, the density increases by about 3% and the CTE changes from $\sim 4\times10^{-6}\,\mathrm{K}^{-1}$ to near zero as shown in Fig. 3.42. During the ceramization of articles of large dimensions, such as flat panels, one has to assure that the crystallization starts simultaneously in each volume element of the article in order to avoid geometrical distortions and to maintain the original shape (except for the linear shrinkage of about 1%). Moreover, one has to take care that the crystallization heat is efficiently removed from the article to avoid spatially inhomogeneous crystallization. In order to meet these

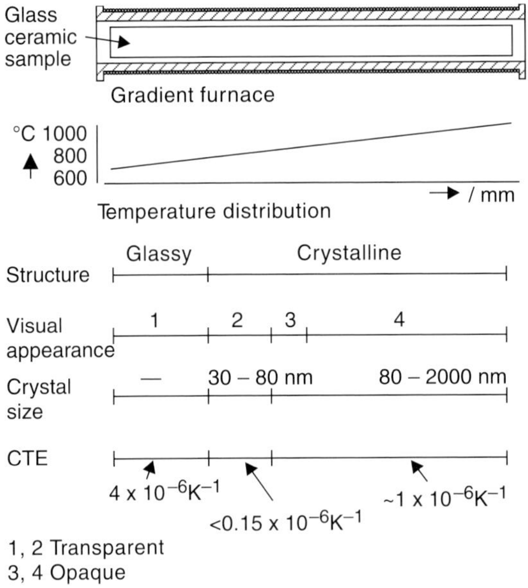

Fig. 3.41. Heat treatment of parent LAS glass in a gradient furnace [3.49]

requirements the actual crystallization takes place at a constant temperature or at least at very low heating rates.

The decoration of cooktop panels is often performed at the same time as their ceramization. In this case the temperature-dependent properties of glazes have to be matched to the ceramization process as well.

The physical and chemical properties of a typical glass ceramic produced in that way and used as a cooktop panel are summarized in Table 3.3 [3.50].

Besides melting, ceramization is the most decisive process stage in manufacturing decorative cooktop panels. During ceramization, the parent glass is converted to glass ceramic using a temperature treatment. It is only at this partial stage that the special properties of the Ceran® cooktop panels,

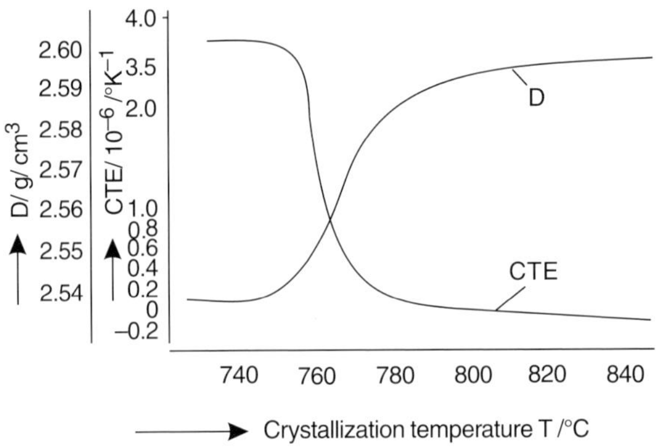

Fig. 3.42. Change of density D and CTE of a glass ceramic during ceramization as a function of crystallization temperature (crystallization time: 2 h) [3.49]

such as the outstanding resistance to temperature shocks and the practically negligible thermal expansion, are produced.

The finished and decorated panels are subjected to a precisely defined time/temperature treatment in a special furnace. The process control system has to ensure not only the setting of the required physical properties of the glass ceramic itself and the decorated areas of the cooktop panel, but also the observance of requirements relating to flatness and homogeneity of the final product. Here the shrinking processes, which occur during ceramization, should be given special consideration with regard to flatness and dimensional stability. Precisely defined temperature gradients must not be exceeded.

3.3.15 Firing of the Ceramic Colors

Extensive trials have demonstrated that the most cost-efficient method, namely firing the decoration colors *during* the ceramization process (see also [3.51]) also results in optimum product properties (see Sect. 3.3 on ceramization). Lengths of green plates are obtained by rolling and cooling the molten glass; the plates are then cut to their final dimensions and edged; printing follows by means of a direct serigraphic method using ceramic color pastes. Following the subsequent drying process, the plates are ceramized in a continuous furnace under special controlled conditions simultaneously with the firing of the decoration colors. The extremely complex solution, diffusion, and rephasing process involved, combined with the continuously optimized chemical composition of the ceramic colors, results in a product that meets all the standard requirements.

3.4 Robax® Transparent Glass Ceramic

Manfred Borens, Torsten Gabelmann, Roland Leroux, Toni Münch

As well as white- and black-tinted glass ceramics mentioned in Sect. 3.1, glass-ceramic suppliers also offer translucent and non-tinted highly transparent glass ceramics on the basis of β-quartz solid solution glass ceramic (Li_2O-Al_2O_3-SiO_2 system). Demanded by the market these glass ceramics show zero expansion in a wide temperature range. They are used for optical (see Chap. 4) and domestic applications like stove and fireplace sight panels and for various window applications, for example fire doors. Further areas of application are cover panels for high performance lights or radiant heaters as well as many other industrial applications.

The most important transparent glass ceramics for domestic applications are called Neoceram™ N-0 (Nippon Electric Glass, Japan) with a coefficient of thermal expansion (CTE) $\alpha_{(20-700\,^\circ\mathrm{C})}$ of approximately $-0.3 \times 10^{-6}\,\mathrm{K}^{-1}$, Keraglass® (Eurokera: joint venture of Corning/St. Gobain, USA/France) with a CTE of approximately $0 \pm 0.2 \times 10^{-6}\,\mathrm{K}^{-1}$, as well as Robax® (Schott AG, Germany).

3.4.1 Robax® Technical Data

The transparent glass ceramic from Schott, called Robax®, has a slight amber shade and its rolled surface exhibits a slight texture from the manufacturing process.

The very low coefficient of linear thermal expansion, its extremely high operating temperature, and excellent resistance to thermal shocks and thermal gradients are outstanding characteristics of Robax®. Since – because of the extremely low thermal expansion – mechanical stresses occurring in the material attributable to changes of temperatures are negligible, the material does not fatigue due to these stresses.

Consequently, this material is suited for applications in which other glasses – including borosilicate glass (e.g. Borofloat®) – fail, due to elevated temperatures, heavy temperature shocks, and wide temperature gradients. In such application fields Robax® frequently proves to be a suitable alternative to quartz glass. Table 3.7 lists some of the Robax® data. For the purpose of comparison, the respective characteristic values for Borofloat® (floated borosilicate glass) are also given there.

The low density and the inherent stability, the high thermal load and the good transparency are characteristics that make this glass ceramic an interesting design material. In many environments, Robax® is chemically resistant.

As compared to metallic or polymeric materials, the properties of Robax® transparent glass ceramic are to be explained by its "vitreous nature". Brittleness and high notch impact strength are of great significance to the user as they make the component's load capacity heavily dependent on the component's surface condition as well as on the edge processing, installation, type of impact, panel size, and thickness, and so on.

Table 3.7. Comparison of some Robax® and Borofloat® glass properties

	Robax®	Borofloat®
Coefficient of mean linear expansion	$0 \pm 0.5 \times 10^{-6}\,\mathrm{K}^{-1}$	$3.25 \times 10^{-6}\,\mathrm{K}^{-1}$
Temperature load capacity	710 °C (10 h)	500 °C (< 10 h)
Short term	560 °C (5000 h)	450 °C (≥ 10 h)
Long term	*)	80 K (> 100) h
Thermal gradients	*)	175 K (th. ≤ 3.8 mm)
Young's modulus	≤ 95 dPa	64 dPa
Poisson's ratio	≈ 0.25	0.2
Density (at 25 °C)	2.6 g/cm^3	2.2 g/cm^3

*) No cracking due to thermal stress at $T_{\mathrm{es,\,max}} \leq 700\,°\mathrm{C}$. $T_{\mathrm{es,\,max}}$ is the maximum temperature on the exterior side of the panel at the hottest point.

3.4.2 Development and Production – Use of Robax® as Stove and Fireplace Sight Panels

Its characteristics make Robax® ideally suited to be used as material for sight panels capable of sustaining high temperatures. Robax® is used for sight and protective panels in combustion furnaces (e.g., for wood, coal, oil, gas), industrial furnaces, and for cover panels in front of "open fireplaces".

Today, flat glass ceramic panels are available in practically all sizes to be used for stoves and fireplaces. Moreover, bent and shaped Robax® panels are available for specific designs, see Chap. 3, Sect. 3.4.3. The current design trend for fireplace sight panels is going towards bigger panel sizes and round bent panels. Furthermore one- and two-angular-bent panels are part of the product range. Highly sophisticated new processes enable the realization of complex 3D shapes with Robax®, for example Robax® Dome (Fig. 3.43 left) and Robax® Elisse (Fig. 3.43 right).

In order to cover constructional elements inside the stove or to offer new and creative design possibilities, Schott developed special ceramic colors for panel decoration. To expand the functionality of the glass ceramic material there are special coatings available, for example oxidic and heat-reflective coatings. The thicknesses most commonly used are 3, 4, and 5 mm.

In practical use, the oven and fireplace sight panels are subjected to stresses mainly by the corrosive effects of the oven atmosphere due to the combustion of, for example, fuel (so-called microcrazing). The intensity of the said stresses is a function of various influential factors.

Corrosive stress may bring about undue damage to the sight panel. This may cause the sight panel to lose its transparency as a result of cloudiness or deposits, or the surface attack may lower the mechanical load capacity to an inadmissibly heavy extent. Typical attack phenomena are deposits or pitting and microcracks. This phenomenon is ascribable to a leaching of the glass ceramic surface (e.g., alkali depletion). Deposits form during the first stage, whereas pitting occurs in the event of continued attack, and due to

Fig. 3.43. Exciting new designs for fireplaces: Robax® Dome (*left*) [3.52] and Robax® Elisse (*right*) [3.53]

the alkali depletion, tensile stresses form in the surface, leading to surface cracks resulting from imperfections. Leaching of the sight panel surface is attributable to aggressive combustion products. It is known that sulfurous waste gases attack the panel in the presence of water (also a combustion product). The analysis of the deposits on oven sight panels (result: high percentage of alkali sulfate) has led to this track.

Surface attack can be avoided or at least reduced by a variety of measures. The simplest way is regular cleaning with an appropriate cleanser. Any deposits, soot, and other dirty substances are removed from the panel surface during such an exercise. Also, the smooth surface, thus produced, effectively delays the formation of deposits. By adding appropriate silicon oil to the cleansing agent, a protective coat is formed on the panel preventing the action of an aggressive oven atmosphere. Protective coatings (e.g. SiO_2 dip coating) markedly reduce surface attack as long as they are intact. Another benefit of coating the panel is that such a coating can assume additional functions (e.g., coatings with a low emission coefficient) in order to reduce radiation from the oven sight panel.

3.4.3 Bending of Robax®

As for Ceran®, also for Robax® bending became more and more popular. Whereas in the early 1990s we registered an increasing demand for angular bent panels, nowadays round bent panels are in a dominating position (Fig. 3.44).

Round bent panels are formed during the ceramization process. The raw glass is cut to size and put on a mold made of fused silica. Before the ceramization starts, the panel's viscosity becomes so high that the panel sags into the mold due to gravity. Depending on the size and the apex angle, additional tools (e.g., ceramic rollers included in the mold) must be used for a smooth bending into the mold. This process allows radii bigger than 200 mm and apex angles up to 180°.

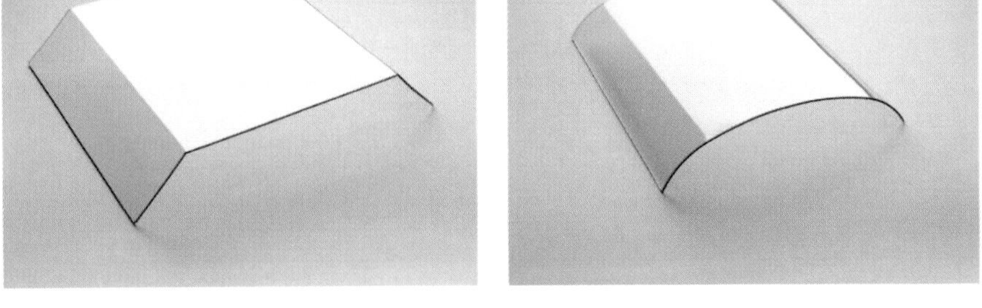

Fig. 3.44. Robax® angular (*left*) and round bent panels (*right*)

As far as angular bent panels are concerned we distinguish between single and double angular bent panels. Production of angular bent panels requires an additional step. After cutting, the glass is “pre-bent” in special equipment. This bending equipment was developed by Schott and is still in progress. The panels are conveyed into a preheating furnace and heated to 500 °C to prevent breakage due to thermal stresses. They are positioned and then transported to the proper bending section. Atmospheric gas burners, which are located above and under the panel, heat the panel rapidly (regionally in the area where the bending edge should be) to more than 1000 °C. The burners are movable vertically and horizontally. By gravitation or with the help of tools the side parts are inclined against a stop unit that can be positioned easily for different glass angles. After bending the panels are conveyed to a reheating section again to reduce thermal stresses. The ceramization process also needs molds to support the “pre-bent” panels. Depending on size and angle, different mold concepts are realized. For more complex shaped panels, a vacuum process or a NIR-based hot-forming process are used (see also Sect. 3.3.12).

3.5 Surface Strengthening of Low Expansion Glass Ceramics

Werner Kiefer

3.5.1 Thermal Toughening of Low-Expansion Glass Ceramics

For glasses, the theoretical strength can be calculated to be of the order of $10^4\ \mathrm{N/mm^2}$. The technical strength of a glass article, however, is only below $10^2\ \mathrm{N/mm^2}$ and depends strongly on the surface of the glass. To improve the technical strength of a glass article it is, therefore, necessary to strengthen the glass surface by producing compressive stresses in the surface. The method most often used for surface strengthening of soda-lime silicate glass is thermal toughening, i.e., the glass is heated up to a temperature between the transformation temperature and the softening point and, afterwards, cooled down rapidly. During the quick cooling-down process, the surface layer will be frozen in a more open structure, i.e., the atomic density of the surface layer is smaller than that in the interior of the glass, because the interior is cooled down more slowly. Therefore, after the cooling process, the surface layer is under compressive stresses and the interior of the glass under tensile stresses [3.54].

The maximum compressive stresses, σ_{max}, developed in this way within a surface layer of a flat glass sheet can be calculated by (3.12), under the assumption that the heat transfer between the surface of the glass and the cooling medium will be ∞, from [3.55]:

$$\sigma_{\mathrm{max}} = \frac{\alpha E}{1-\mu}(T_{\mathrm{g}} - T_{\infty}) \ , \tag{3.12}$$

where α is the coefficient of thermal expansion (CTE), E is Young's modulus, μ is Poisson's ratio, T_g is the transformation temperature, and T_∞ is the temperature of the cooling medium.

In most cases, low-expansion glass ceramics are not completely crystallized and have, therefore, a certain amount of a residual glass phase. Due to this vitreous phase there is also a transformation temperature for the crystallized glass ceramics. The transformation temperature of crystallized glass ceramics is always higher than the transformation temperature of non-crystallized glass ceramics. The Schott glass ceramics of type number 8562 have a T_g of 670 °C for the vitreous state before ceramization and a T_g of 730 °C for the ceramized glass ceramics (see Fig. 3.45).

From (3.12) it can be derived that it is usually not reasonable to thermally toughen low-expansion glass ceramics. Glass ceramics for technical applications may have a thermal expansion coefficient (CTE) of up to $0.3 \times 10^{-6}\,\mathrm{K}^{-1}$. With a Young's modulus of $E = 90 \times 10^3$ MPa, $\mu = 0.25$, $T_g = 730\,°\mathrm{C}$, and $T_\infty = 30\,°\mathrm{C}$, maximum compressive stresses of 25 MPa can be calculated by (3.12). This amount is about half of the strength of the non-toughened glass ceramics. Thus, the whole strength which could be obtained in this case is between 75 and 100 MPa. The amount of 25 MPa of compressive stresses in the surface layer is not high enough for the glass sheet to disintegrate into small crumbs like safety glass, which is used to provide protection in the case of accidental human impact.

Low-expansion glass ceramics, however, show excellent thermal shock resistance and high thermal load capacity. For that reason they are used as table

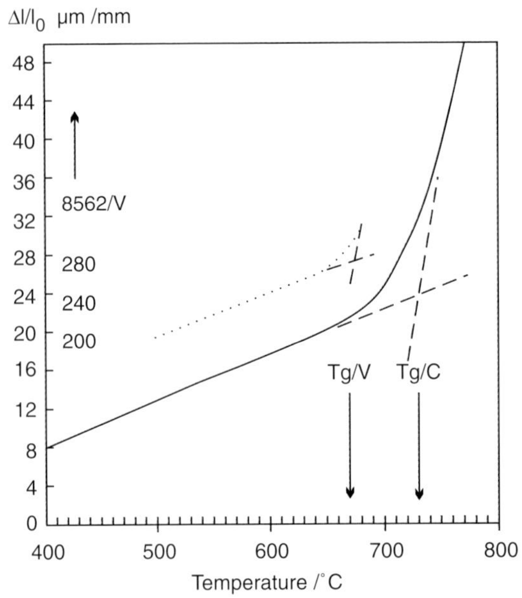

Fig. 3.45. Transformation temperature T_g and thermal expansion $\Delta l/l_0$ of the glass ceramic 8562 (V = in the vitreous state, C = in the ceramized state)

and oven ware, oven plates, or fire resistant glasses. With all these applications, low-expansion glass ceramics are exposed to relatively high temperatures. In most of these cases, compressive stresses due to thermal toughening would be reduced.

The low compressive stresses obtainable by thermal toughening and the reduction of prestressing in daily use are the reasons why low-expansion glass ceramics are usually not prestressed by thermal toughening.

3.5.2 Chemical Strengthening of Low-Expansion Glass Ceramics

With glasses chemical strengthening can be performed by an exchange of ions for larger ions below T_g or smaller ions above T_g.

For low-expansion glass ceramics it is appropriate to subdivide into (a) chemical strengthening of crystallized glass ceramics with larger ions below T_g, (b) chemical strengthening of glass ceramics by Li exchange before crystallization, and (c) chemical strengthening of crystallized glass ceramics by Li exchange.

Chemical Strengthening of Crystallized Glass Ceramics with Large Ions Below T_g

For the chemical strengthening of glasses or crystallized glass ceramics below T_g, small ions from the surface layer are exchanged for larger ions, as Li^+ ions for Na^+ ions or Li^+ and Na^+ ions for K^+ and Cs^+ ions from a molten salt bath. By this ion exchange, the structure in the surface layer does not change, but it is stuffed with the larger ions. A compressive stress is developed during cooling down [3.56]. Normally, ion exchange in glasses takes place in a molten potassium nitrate (KNO_3) salt bath because KNO_3 has a low melting temperature.

Low-expansion glass ceramics have a crystalline as well as a vitreous phase. In this vitreous phase normal ion exchange can take place. But for the chemical strengthening of low-expansion glass ceramics at transformation temperatures above 700 °C, salt melts of potassium nitrate cannot be used, because the salt melts of nitrates decompose at temperatures above 500 °C and produce the toxic NO_x. Other potassium salts such as sulfates and chlorides and their eutectics have melting temperatures above 700 °C.

Therefore, special potassium pastes have been developed which do not decompose. They consist of a potassium salt, such as potassium sulfate (K_2SO_4), or a mixture of potassium sulfate and chloride (e.g., 52% in weight K_2SO_4 + 48% in weight KCl) and different organic solvents [3.57].

The surfaces of the glass articles can be coated with these potassium pastes in a spraying or dipping process. After drying, the articles are heated up to the ion exchange temperature. There is a weak sintering of the paste during the ion exchange. After the ion exchange the glass articles are cooled down and the paste can easily be removed from the surface without any visible residues.

Ion exchange experiments were made with Schott glass ceramics 8562 developed for cookware. Crystallized glass ceramics 8562 show a transformation temperature of 730 °C. The best ion exchange temperature proved to be 700 ± 5 °C. After 4 h a 50 μm deep ion exchange layer can be produced, whereas after 16 h the layer is 100 μm deep (see Fig. 3.46). After abrasion a bending strength of 230 ± 20 MPa was measured, in comparison to 60 MPa of the unstrengthened glass ceramics.

Figure 3.47 shows the maximum temperature and time, where the chemically strengthened articles of glass ceramics 8562 can be used without loss of compressive stresses.

Chemical Strengthening of Glass Ceramics by Li Exchange Before Crystallization

During the chemical strengthening of glasses above T_g, large Na^+ or K^+ ions from the glass surface layer are exchanged for smaller ions, Li^+ and Na^+ ions, respectively, from a molten salt bath or a paste. This ion exchange causes a change of the surface structure. The exchanged surface layer has a lower thermal expansion coefficient (CTE) than the interior of the glass.

When the glass article cools down, a compressive stress is developed due to that lower CTE of the surface layer. The compressive stress, which can be produced between the surface and the interior of the glass by the ion exchange above T_g, is considerably smaller than that obtained by ion exchange below T_g [3.58]. Therefore, this kind of ion exchange is usually not applied for strengthening glass articles.

In lithium-alumo-silicate glass ceramics a higher lithium concentration due to ion exchange is reached within the surface layer than within the glass. This leads to a higher content of crystal phase in this exchanged layer during the ceramization, because lithium is incorporated into the crystal layer.

If the lithium exchange takes place above the transformation temperature of the vitreous glass ceramics, but below the crystallization temperature, a thin crystalline surface layer of a thickness of a few μm is generated which is under high compressive stress. This thin crystalline layer does not change its thickness during the following ceramization process. The same results will be obtained if the ion exchange takes place during the ceramization. The crystalline surface layer shows a thickness of only about 5 μm and compressive stresses in both cases between 150 and 250 MPa in a layer thickness of 20–30 μm.

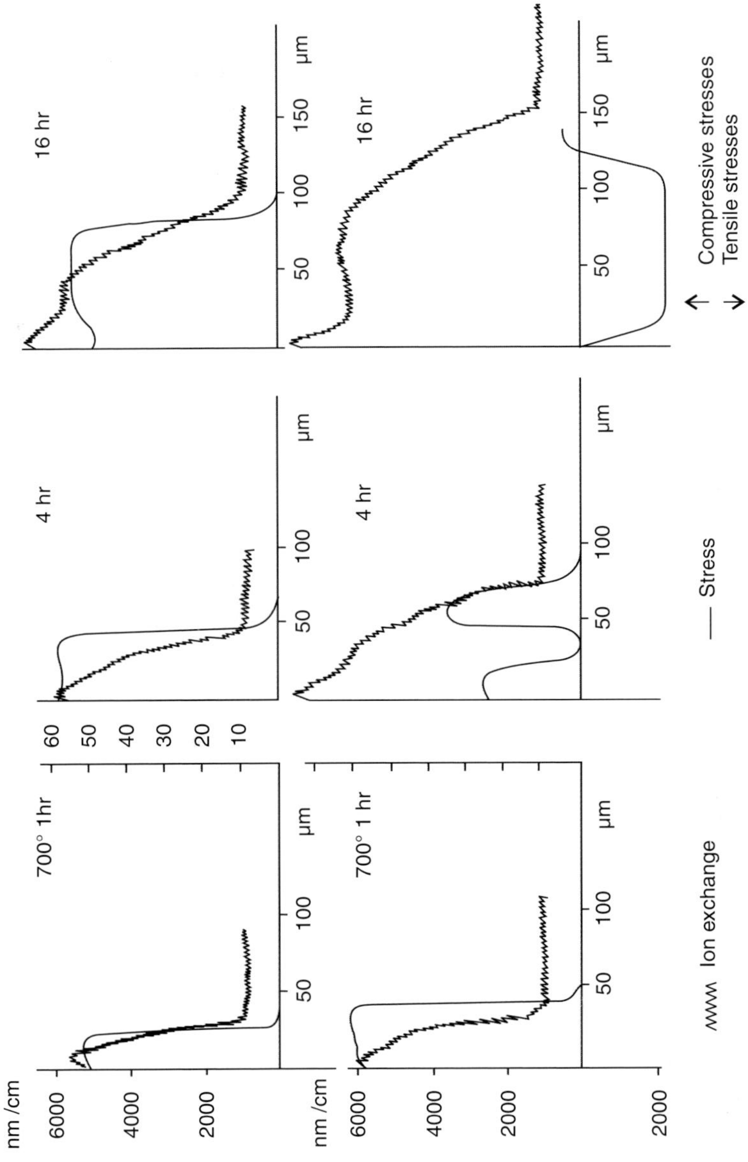

Fig. 3.46. Comparison of ion exchange layer and stress distribution in dependence on tempering time and temperature

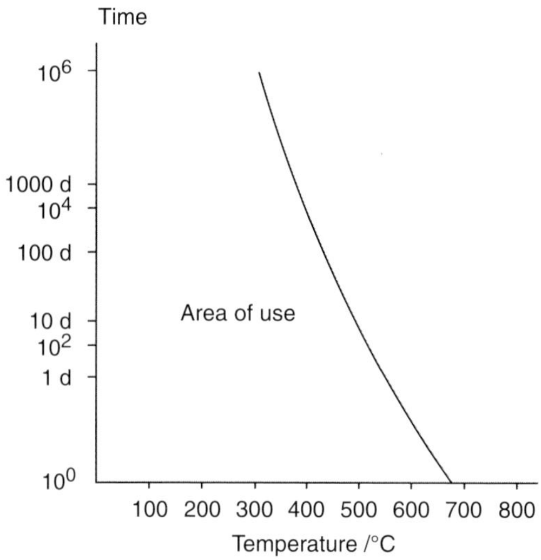

Fig. 3.47. Maximum temperature and time of use for glass ceramics 8562 after ion exchange with potassium

Chemical Strengthening of Crystallized Glass Ceramics by Li Exchange

The exchange of ions in a ceramized glass ceramic 8562 for the smaller Li ions from molten salts always produces compressive stresses, regardless of whether the exchange takes place below or above the transformation temperature. With rising exchange time and temperature, the depth of the compressive stress layer increases very slowly and attains a depth of 50–70 μm after 16 h at 760 °C or after 4 h at 800 °C. The compressive stresses obtained between 100 and 150 MPa are nearly independent of the temperature.

Lithium ion exchange is rarely used in practical applications for prestressing lithium-alumo-silicate glass ceramics. On the one hand, the compressive stresses are lower in comparison to the ion exchange with potassium below T_g, and on the other hand, most Li salts attack the surface so that the articles are no longer usable.

3.5.3 Surface Strengthening by Surface Crystallization

In case of surface strengthening by surface crystallization a crystalline surface layer is generated. This crystalline surface layer shows a distinctly lower thermal expansion coefficient (CTE) than the base glass. During cooling down of such a surface-crystallized article, the interior of the glass with the higher CTE contracts more than the surface layer with the lower CTE. After cooling down, the surface layer is under compressive stresses (σ_c) and the interior is under tensile stress (σ_t).

The compressive stresses in the surface (σ_c) and the tensile stress in the interior (σ_t) can be roughly calculated by (3.13) and (3.14),

$$\sigma_{\mathrm{c}} = \frac{E}{1-\mu}\,\Delta\alpha\,T_{\mathrm{g}}(1 - d_{\mathrm{cr}}/d_{\mathrm{gl}}) \tag{3.13}$$

$$\sigma_{\mathrm{T}} = \frac{E}{1-\mu}\,\Delta\alpha\,T_{\mathrm{g}}(d_{\mathrm{cr}}/d_{\mathrm{gl}})\ , \tag{3.14}$$

where $\Delta\alpha$ is the difference of CTE between the crystalline surface layer and the interior of glass, T_g is the transformation temperature of the base glass, d_{cr} is the thickness of the crystalline surface layer, and d_{gl} is half the thickness of the glass. To simplify matters, E_{cr} was equated with E_{gl} and μ_{cr} with μ_{gl}.

In most cases, the objective is to strengthen glass articles that are as thin as possible at high temperatures. However, the tensile stresses within the glass must be prevented from getting too high, as otherwise self-destructions may easily occur. Values gathered from field applications show that tensile stresses within the glass should not exceed 50–70 MPa.

The thickness of the crystalline layer should be 100–150 μm so that the cracks caused by scratching in daily use, which may reach a depth of 30–50 μm, do not extend the area of compressive stresses.

As base glasses for strengthening by surface crystallization, primarily low-expansion glass ceramics are eligible because after ceramization they exhibit a lower thermal expansion than does the base glass. The difference in thermal expansion is between 4 and 5×10^{-6} units.

For investigations on surface crystallization a glass ceramics composition close to the composition of Schott glass ceramics 8562 was chosen. Volume crystallization was prevented by taking out the nucleating agents. Crystallization starts in defects on the glass surface, and the crystals grow from there towards the center of the glass. The crystal growth rate is constant, i.e., directly proportional to time. Crystal growth is thus not directly diffusion-controlled. Depending on the treatment of the surface, crystallization starts with a certain delay (see Fig. 3.48).

Nucleation, and thus commencement of crystallization, can be accelerated by coating with TiO_2 dipping layers and applying lithium salt pastes and water vapor (see Fig. 3.49).

To keep any deformation of the glass articles during the surface crystallization to a strict minimum, we must try to obtain a compact crystalline layer at temperatures as low as possible.

It became apparent that the crystal growth rate can be accelerated by adding Na_2O to the glass composition, although Na^+ is not incorporated into the crystal. By investigations with a microprobe it was possible to prove that with Li^+, Na^+ performs an internal ion exchange in the glass. Na^+ located on the crystal front migrates to the center of the glass, and Li^+ migrates from the center of the glass to the crystal front where it is incorporated into the crystallized layer (see Fig. 3.50 A).

Potassium is not incorporated into the crystal. The frontier of the crystallization pushes the potassium ahead, as is also clearly visible from the microprobe depth profile (see Fig. 3.50 B). The diffusion rate of the bivalent

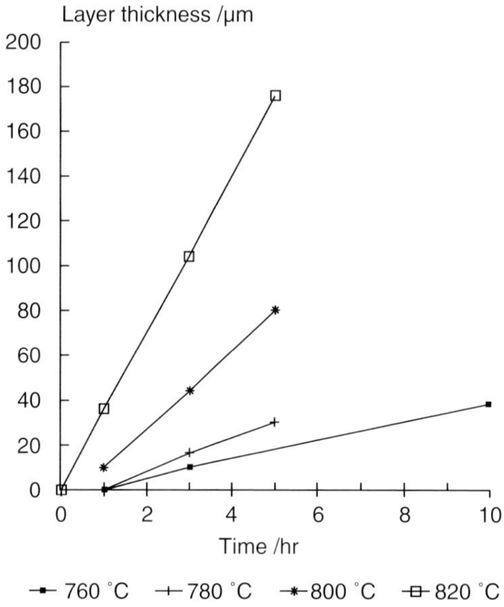

Fig. 3.48. Thickness of a crystalline surface layer in dependence of tempering time for different temperatures

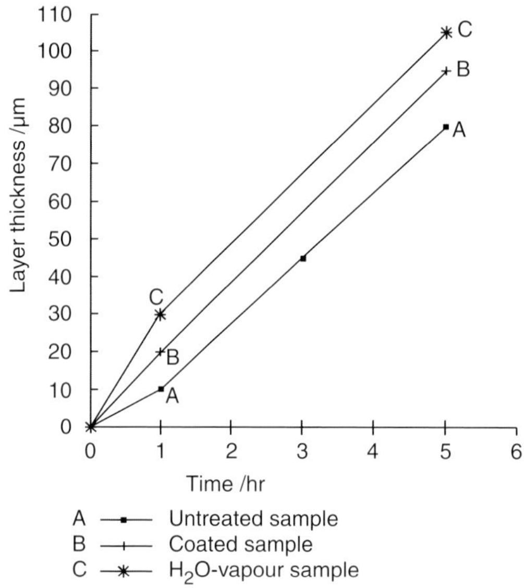

Fig. 3.49. Thickness of a crystalline surface layer in dependence of tempering time for different treatments: (A) untreated, (B) coated with TiO_2, (C) tempered in a water vapor atmosphere

ions that cannot be incorporated into the crystal lattice, such as Ca^{2+} and Ba^{2+}, is also pushed ahead by the crystal front; but, thereafter, the crystal grows around this residual glass phase. Under the microscope, the crystals are of a fern-like appearance (see Fig. 3.50 C) [3.59].

With an increasing Na_2O content in the glass composition, the crystals growing from the glass surface into the glass become more and more pointed.

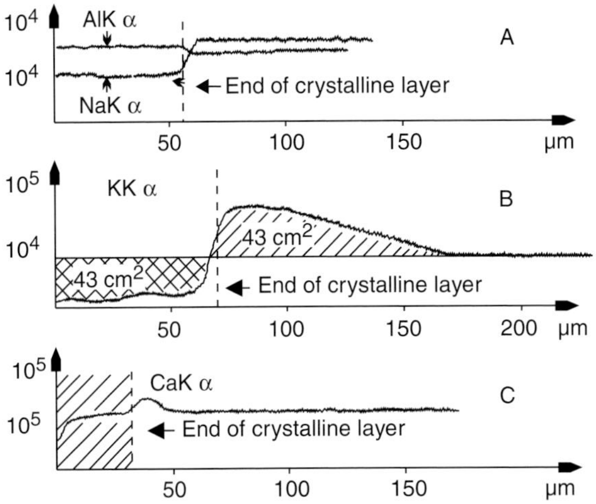

Fig. 3.50. Microprobe measurement of the internal ion exchange between the crystalline surface layer and the interior of the glass for (A) Na^+ and Al^{3+}, (B) K^+, and (C) Ca^{2+}

Excessive tensile stresses which easily lead to cracks are generated on these crystal tips. In surface crystallization the crystal front is not to show any tips, but only rounded-off crystals. This can be controlled primarily via the glass composition. Thus the addition of B_2O_3 leads to beautifully rounded-off crystals [3.59].

With the composition of the base glasses and in consideration of the internal ion exchange in the surface crystallization, large α-value differences between the glass surface and glass center can be generated. An external ion exchange is less suitable for this as, due to the crystalline surface layer, no ion exchange with the center of the glass takes place. The crystalline surface layer acts as a barrier layer.

Investigations have shown that a too-large α-value difference results in the destruction of the test specimens during the cooling-down process. If the product $\Delta\alpha\,T_g$ exceeds the value of 4.4×10^{-3}, the specimens shatter into pieces. A rough calculation according to (3.13) and (3.14) exemplifies this and shows the limits of the process.

For $\Delta\alpha = 6.0 \times 10^{-6}$, $T_g = 650\,°C$, $E = 90 \times 10^3$ MPa, $\mu = 0.25$, $d_{cr} = 100\,\mu m$, and $d_{gl} = 1000\,\mu m$,

$$\sigma_c = \frac{90 \times 10^3}{0.75} 6 \times 10^{-6} \times 650(1 - 100/1000) = 412 \text{ MPa} ,$$

$$\sigma_t = 468(100/1000) = 46.8 \text{ MPa} . \tag{3.15}$$

It is a well-known fact from the manufacture of safety glasses that tensile stresses within the glass should not exceed 50–70 MPa, as otherwise the risk of self-destruction becomes too great.

Although compressive stresses of 300–500 MPa can be obtained via the process of surface crystallization, this method is not applied in practical use. For one thing, it is annoying that, due to the crystalline surface layer, the articles are no longer clear and transparent, but only translucent. It is, however, much more aggravating that it is hardly possible to manufacture articles not containing any defects or contaminations in their interior, like, for example, platinum particles acting as nucleating agents just like the defects on the surface. From these nuclei within the glass, crystal growth starts likewise. Crystals in the center of the glass produce cracks in the layer under tensile stresses, which result in a delayed destruction of the strengthened articles (see Fig. 3.51). In addition to the excessive stresses before the crystal front, the crystals inside the glass are thus the main reason why strengthening by surface crystallization has not gained acceptance in practical applications.

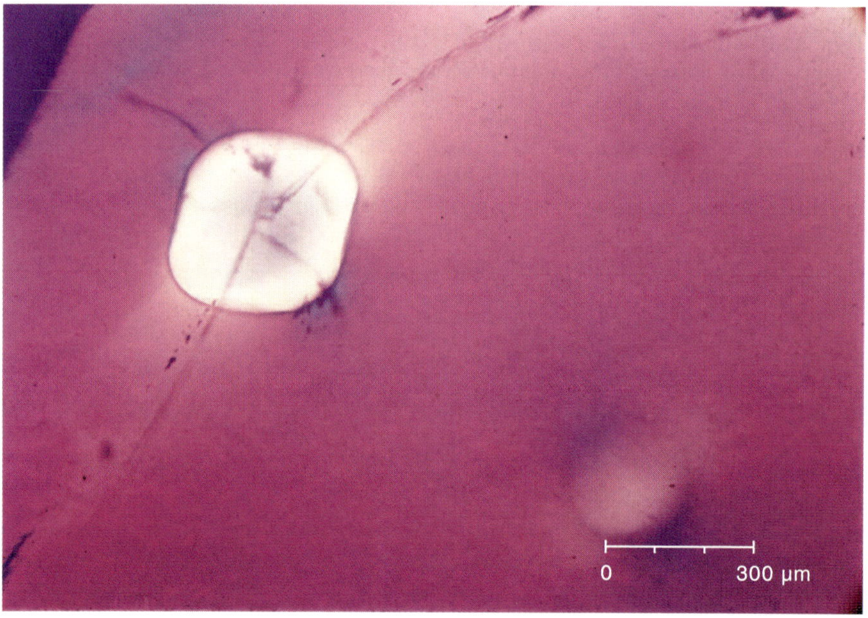

Fig. 3.51. Photograph of a crystal with cracks in the interior of a glass piece with a crystalline surface

References

3.1 W. Höland, G. Beall: Glass-Ceramic Technology, XV ff., 75 ff., 247 ff. (Am. Ceram. Soc., Westerville OH 2002)
3.2 Schott Glaswerke: "Ceran, Die Geschichte eines Produkterfolges" (Mainz 1993)

3.3 Reprinted with permission from Rinnai Corporation, Nagoya, Japan
3.4 S.P. Timoshenko, I.N. Goodier: *Theory of Elasticity* (McGraw-Hill, New York 1982) pp. 441–443
3.5 B. Wölzlein: "Experimentelle Bestimmung der Lage der Zugspannungsmaxima bei partieller Beheizung von Glas- und Glaskeramikplatten bei Variation des Abstandes zwischen beheizter Fläche und kaltem Rand" ("Experimental determination of the location of tensile-stress maxima with partial heating of glass and glass ceramic plates, and with variation of the distance between heated surface and cold edge") Diploma Thesis (Georg-Simon-Ohm Fachhochschule Nürnberg, Fachbereich Werkstofftechnik, Nurenberg, July 31, 1984)
3.6 A.Q. Tool: "Relexation between inelastic deformability and thermal expansion of glass in its annealing range", J. Am. Ceram. Soc. **29**(9), 240–53 (1946)
3.7 O.S. Narayanaswamy: "A model of structural relaxation in glass", J. Am. Ceram. Soc. **54**(10), 491–498 (1971)
3.8 H. Scholze: *Glas. Natur, Struktur und Eigenschaften*, 3rd ed. (Springer, Berlin, Heidelberg 1988)
3.9 F. Kerkhof: *Bruchvorgänge in Gläsern* (Dtsch. Glastechnische Gesesllschaft, Frankfurt/Main 1970)
3.10 G. Exner: "Erlaubte Biegespannung in Glasbauteilen im Dauerlastfall", Glastechn. Ber. **56**(11), 299-312 (1983)
3.11 N. Kamiya, O. Kamigaito: "Application of fracture mechanics to heterogeneous systems – prediction of fatigue life of ceramics", J. Mat. Sci. **19**(1), 159–66 (1984)
3.12 K.S. Chan, R.A. Page: "Creep-crack growth by damage accumulation in a glass ceramic", J. Am. Ceram. Soc. **74**(7), 1605–1613(1991)
3.13 B.S. Majumdar, T. Mah, M.G. Mendiratta: "Flaw growth in a polycrystalline lithium-aluminium-silicate glass ceramic", J. Mat. Sci. **17**(11), 3129–3139 (1982)
3.14 R.K. Govila, K.R. Kinsman, P. Beardmore: "Fracture phenomenology of a lithium-aluminium-silicate glass-ceramic", J. Mat. Sci. **13**(10), 2081–2091 (1978)
3.15 J.P. Northover, G.W. Groves: "High-temperature mechanical properties of Li2O-Al2O3-SiO2 (LAS) glass ceramics. II. Fracture", J. Mat. Sci. **16**(7), 1881–1886 (1981)
3.16 M. Tomozana: "High-toughness glass-ceramics", Report (1994) ARO-28577, Gov. Rep. Announce. Index (U.S.) 1995, 95(19)
3.17 G.L. Leatherman, M. Tomozawa: "Mechanical properties of a transformation-toughened glass-ceramic", J. Mat. Sci. **25**(10) 4488–4494 (1990)
3.18 LMBG (Lebensmittel- und Bedarfsgegenständegesetz), August 15, 1974
3.19 S.D. Stookey: "Method of making ceramics and product thereof", US Patent 2,920,971, January 12, 1960
3.20 H. Scheidler, E. Rodek: "Li_2O-Al_2O_3-SiO_2 glass ceramics", Am. Ceram. Bull. **68** (Nov. 1989)
3.21 P.W. McMillan: *Glass-Ceramics* (Academic Press, London, New York 1964)
3.22 D.R. Stewart: "Glass-ceramics", in *Encyclopedia of Glass, Ceramics, Clay and Cement*, ed. by M. Grayson (Wiley, New York 1985) pp. 538–549
3.23 G.H. Beall, D.A. Duke: "Glass-ceramic technology", in *Glass: Science and Technology, Vol. 1: Glass-Forming Systems*, ed. by D.R. Uhlmann, N.J. Kreidl (Academic Press, New York 1983) pp. 403–445
3.24 V. Maier, G. Müller: "Mechanism of oxide nucleation in lithium aluminosilicate glass-ceramics", J. Am. Ceram. Soc. **70**, C176–C178 (1987)

3.25 P.E. Doherty, D.W. Lee, R.S. Davis: "Direct observation of the crystallization of Li_2O-Al_2O_3-SiO_2 glasses containing TiO_2", J. Am. Ceram. Soc. **50**, 77–81 (1967)
3.26 H. Scheidler, W. Sack: "The various effects of the nucleating agents ZrO_2 and TiO_2 on the crystallization behavior of some lithium-aluminum-silicate (glass-ceramic) glasses", in *IX. Int. Congress on Glass* (Paris 1971) pp. 1069–1085
3.27 G. Müller: "Solid-solution problems in glass-ceramic systems", Fortsch. Mineral. **52**, 182–192 (1975)
3.28 U. Schiffner, W. Pannhorst: "Nucleation in a precursor glass for a Li_2O-Al_2O_3-SiO_2 glass ceramic. Part 1: nucleation kinetics", Glastechn. Ber. **60**, 211–221 (1987)
3.29 U. Schiffner, W. Pannhorst: "Nucleation in a precursor glass for a Li_2O-Al_2O_3-SiO_2 glass ceramic. Part 2: variations of the nucleating agent concentrations", Glastechn. Ber. **60**, 239–247 (1987)
3.30 Schott Glas: "All in white", Schott Information **97**, 18–19 (I/2001)
3.31 M.J. Buerger: "The stuffed derivates of the silica structures", Am. Miner. **39**, 600–614 (1954)
3.32 J. Petzoldt: "Untersuchungen an transparenten, metastabilen Quarz-Mischkristalle enthaltenden Glaskeramiken", Glastechn. Ber. **43**, 127–137 (1970)
3.33 J. Petzoldt, W. Pannhorst: "Chemistry and structure of glass-ceramic materials for high precision optical applications", J. Non-Cryst. Solids **129**, 191–198 (1991)
3.34 C.L. Babcock, R.A. Busdiecker, E.C. Hagedorn: "Method of making low expansion crystallized glass-ceramics and telescope mirror blanks made thereby", US Patent 4,285,728, August 25, 1981
3.35 K. Chyung: "Transparent beta-quartz glass-ceramics", US Patent 4,018,612, April 19, 1977
3.36 H.L. Rittler: "Colored transparent, translucent and opaque glass-ceramics", US Patent 4,461,839, July 24, 1984
3.37 T. Shibuya, K. Matsui, M. Matsumoto: "Infrared transparent glass ceramic articles with betaquartz solid solution crystals without any other crystals", US Patent 4,835,121, May 30, 1989
3.38 W. Pannhorst, E. Rodek, H. Scheidler: "Transparent colored glass ceramic with good thermal stability and variable transmission in the IR region", US Patent 5,212,122, May 18, 1993
3.39 Dietzel: Mitt. VDEfa 10 (1962) p. 35
3.40 E. Rodek, W. Kiefer, F. Siebers: "Verwendung von blei- und cadmiumfreien Glaszusammensetzungen zum Glasieren, Emaillieren und Verzieren und deren Zusammensetzung", German Patent DE 4 201 286 C2
3.41 W. Pannhorst, E. Rodek, H. Scheidler: "Durchsichtige farbige Glaskeramik mit guter Temperaturbelastbarkeit und variabel einstellbarer Transmission im IR-Bereich", European Patent 0 220 333, October 26, 1985
3.42 H.A. Schaeffer: *Technologie des Glases* (Institut für Werkstoffwissenschaften III, Universität Erlangen-Nürnberg 1990) pp. 44–47
3.43 W. Trier: *Glasschmelzöfen – Konstruktion und Betriebsverhalten* (Springer, Berlin, Heidelberg 1984) p. 63
3.44 G.L. Moilanen, B. v. Kalsbeek: "NOx control options for glass furnaces", Ceram. Eng. Sci. Proc. **12**, 632–649 (1991)
3.45 H. Jebsen-Marwedel, R. Brückner (Eds.): *Glastechnische Fabrikationsfehler* (Springer, Berlin, Heidelberg 1980) pp. 193–268
3.46 W. Trier: "Berechnung der Temperaturverteilung und Wärmeströme in der Walze", Glastechn. Ber. **38**(11), 438–450 (1965)
3.47 Reprint with permission from Schott Instruments

3.48 Reprint with permission from BSH Bosch und Siemens Hausgeräte GmbH, Traunreut, Germany
3.49 H. Scheidler: "Herstellung und Eigenschaften von Glaskeramik-Werkstoffen", Silikat-Journal **11**, 144–149 (1972)
3.50 H. Scheidler, M. Taplan: "Ceran glass ceramic hobtops", Schott Information **2**, 3–21 (1984)
3.51 DE 3505922-C: "Verfahren zum Dekorieren von Glaskeramik-Platten"
3.52 Reprinted with permission from Spartherm
3.53 Reprinted with permission from Spartherm
3.54 W.D. Kingery, H.K. Bowen, D.R. Uhlmann: *Introduction to Ceramics*, 2nd ed. (Wiley, New York 1975) pp. 830–833
3.55 W. Kiefer: "Thermisches Vorspannen von Gläsern niedriger Wärmeausdehnung", Glastechn. Ber. **57**, 221–228 (1984)
3.56 W.D. Kingery, H.K. Bowen, D.R. Uhlmann: *Introduction to Ceramics*, 2nd ed. (Wiley, New York 1975) pp. 841–844
3.57 W. Kiefer, E. Rodek: "Verfahren zum Ionenaustausch an Glas oder Glaskeramik mittels einer Salzschicht", German Patent D 3 840 071
3.58 W.D. Kingery, H.K. Bowen, D.R. Uhlmann: *Introduction to Ceramics,* 2nd ed. (Wiley, New York 1975) p. 843
3.59 W. Kiefer: "Verfahren zur Herstellung von hochfesten und temperaturwechselbeständigen Glasgegenständen durch Oberflächenkristallisation", German Patent D 2 263 234

4. Zerodur® – A Low Thermal Expansion Glass Ceramic for Optical Precision Applications

4.1 Development of the Optical Glass Ceramic Zerodur®

Wolfgang Pannhorst

When in 1957 Corning Glass Works announced the invention of a new class of materials called glass ceramics, Schott started research in this field to acquire competence for future product developments. Schott mainly concentrated its research on glass ceramics in the Li_2O-Al_2O_3-SiO_2-composition field and contributed to a broadened understanding of the nucleation and crystallization phenomena; for example, *Sack* and *Scheidler* [4.1] were the first to propose the use of a combination of the nucleating agents TiO_2 and ZrO_2. In these early stages of research and development, product ideas mainly centered around cookware. In discussions during international meetings with scientists from other laboratories, Schott scientists learned that the low-expansion Li_2O-Al_2O_3-SiO_2 (LAS) glass ceramics were considered as substrate materials for telescope mirror blanks, and perhaps they dreamed of developing such a material. But Schott was well aware that the development of a telescope mirror substrate material and its realization in the form of large castings depended not only on the corresponding demand but equally on the willingness of an institution to supply sufficient funding.

4.1.1 Laboratory Development

The situation changed suddenly when Dr. Elsässer from the Max Planck Institute (MPI) of Astronomy in Heidelberg approached Schott in 1966 asking whether Schott thought that large castings with about 4 m in diameter could really be fabricated from the new low-expansion glass ceramic and whether Schott would be willing to start a corresponding development and fabrication project. Schott declared that it was prepared to evaluate the proposal in more detail; it started a materials development project under the leadership of Dr. Petzoldt while further details of the future project were discussed and planned. The following requirements served as the main guidelines for the laboratory development.

(1) The coefficient of thermal expansion (CTE) should be as close to zero as possible in the temperature interval from −30 °C to +70 °C. It should be about an order of magnitude smaller than the value of 0.43×10^{-6}/K of fused silica for the same temperature interval.
(2) A mirror blank should maintain its shape at a given temperature after many cycles of temperature changes. The specific Young's modulus should be high so that the thickness necessary for the substrate not to deform under gravity forces would be small; the values to compare it with are those for borosilicate glass ($E = 63$ GPa, $\rho = 2.22 \times 10^3$ kg/m^3) and fused silica ($E = 72$ GPa, $\rho = 2.20 \times 10^3$ kg/m^3), materials which serve successfully as telescope mirror substrates.
(3) The material should be homogeneous with respect to all chemical and physical properties. The requirements for inclusions and bubbles should agree with the requirements for optical glasses.
(4) Quality control should be possible by standard optical methods; therefore, the material has to be sufficiently transparent in the visible range, also for large thicknesses in the range of 0.5 m.
(5) Optical polishing should be possible down to accuracies of about $\lambda/10$–$\lambda/20$ for $\lambda = 546$ nm; i.e., the size of the crystallites and the hardness differences between the residual glass and the crystallites should be as small as possible.
(6) The adhesion of an aluminum coating on the polished surface should be as good as for competing materials.
(7) As aluminum coatings are replaced many times during the life-cycle of a telescope, because their reflectivity degrades with time, the chemical resistance of the residual glass and crystalline phase to cleaning operations which have to be designed should be comparable and high. The cleaning operations should have no influence on the figuring of the mirror so that a new coating could be applied without any additional polishing procedure, as it would be impossible to perform the latter at the site of the telescope.

These requirements had in part to be transformed into physical and/or chemical properties of the glass and/or glass ceramic. From the beginning it was requirement 3 that produced the main worries. Would it be possible to achieve sufficient homogeneity with a glass that was very long and had viscosity values as high as about 500 dPa s at 1600 °C, i.e., the temperature limit for conventional glass melting?

Requirements 1 and 2 would be the direct result of the material development and Dr. Petzoldt and his team were confident that good values were achievable. Requirement 4 looked tough; but due to the work of *Sack* and *Scheidler* [4.1] and *Tashiro* and *Takagi* [4.2], who had shown that the yellow color in the LAS glass ceramic can be reduced by using ZrO_2+TiO_2 as nucleating agents instead of TiO_2 alone, it was hoped that by additional work the transparency could be further improved. There were some data available on requirements 5–7 but research was needed to find out whether these requirements could be met. The least could be said about requirement 5, but, fortunately, this requirement could be tested on small samples, so that the laboratory development would provide the answers.

Certainly, the most intriguing requirement was requirement 3. All the others could be tested on samples from laboratory melts; the requirement for homogeneity, however, fully depended on the melting and ceramization technology. So, this requirement had to be transformed into property requirements measurable on laboratory samples. As there was no previous experience with such a transformation it had to be performed on the basis of good guesses by well-experienced people. From all that was known it was clear that the base glass would have a viscosity characteristic far different from any optical glass; it would be rather similar to that of the borosilicate glass Duran®, which is a long glass with relatively high viscosity values even at 1500–1600 °C, but with the viscosity curve shifted to even higher temperatures. The requirements for the ceramization process were even less well-known. The ideal conditions would be that all volume elements would pass through the same time–temperature history curve during cooling of the base glass to room temperature and during ceramization. This, of course, is not possible; so the question was, what gradients would be allowed? Is there a plateau of properties with respect to the time–temperature history? How broad can it be made by varying the composition?

It was these questions that permanently occupied Dr. Petzoldt's mind while he investigated various compositions to extend the possible composition field for the CTE requirement as far as possible. First he investigated the pseudo-quarternary composition field $Li_2Al_2O_4$-$MgAl_2O_4$-$ZnAl_2O_4$-SiO_2 [4.3] whose subsystems $Li_2Al_2O_4$-SiO_2 (e.g., *Saalfeld*, [4.4]) and $MgAl_2O_4$-SiO_2 [4.5] had already been investigated rather thoroughly. He was able to show that h-quartz solid solution (h-quartz s.s.) crystals with low expansion also form in the pseudo-binary system $ZnAl_2O_4$-SiO_2 if the ceramization conditions are well chosen. The h-quartz s.s. crystals are not very stable and transform easily into Zn-gahnite and quartz. On the basis of these findings he was able to extend the field of the formation of metastable h-quartz s.s. crystals within the pseudo-quarternary system considerably. It became clear that starting from an arbitrary LAS composition and increasing the $Li_2Al_2O_4$ amount influences the CTE strongly into the negative direction, additions of $ZnAl_2O_4$ also reduce the CTE value, while additions of $MgAl_2O_4$ increase the CTE.

In a further effort, Petzoldt investigated the possibility of substituting part of the SiO_2 in the metastable h-quartz s.s. crystals by $AlPO_4$. It was known that SiO_2 as well as $AlPO_4$ crystallize into the three polymorphs quartz, crystobalite, and tridymite. Under normal pressure no solid solution between $AlPO_4$ and SiO_2 is found for the quartz polymorphs [4.6].

But as in this case the solid solution formation of $AlPO_4$ in metastable h-quartz crystals of the polynary system Li_2O-MgO-ZnO-Al_2O_3-SiO_2 was of interest, the observations of the pure $AlPO_4$-SiO_2 system could be taken only as an indicator for the chances to substitute part of the SiO_2 by $AlPO_4$; definite evidence could be obtained only by experiments.

Petzoldt investigated a number of pseudo-binary systems such as $Li_2O\cdot Al_2O_3\cdot 2SiO_2$-$AlPO_4$ or $15Li_2O\cdot Al_2O_3\cdot 15ZnO\cdot Al_2O_3\cdot 70SiO_2$-$AlPO_4$ [4.7]. By melting glasses with additions of the nucleating agents ZrO_2 and TiO_2 and choosing suitable ceramization conditions he was able to show that the glasses transformed into h-quartz s.s. containing glass ceramics when up to 50–70 wt% SiO_2 was substituted by $AlPO_4$; the upper limit of the substitutions varied from one system to the other. The CTE values of the glass ceramics obtained depended of course on the ceramization condition chosen, but in many cases they were below $1 \times 10^{-6}\,K^{-1}$, thereby indicating that these compositions could be considered for further evaluation in selecting the final composition. As an example of the investigations performed, Fig. 4.1 displays the variation of the lattice constants of the h-quartz s.s. crystals and of the CTE (20/300) values of the glass ceramics after ceramization at 850 °C for 2 h for the system $20Li_2O\cdot Al_2O_3\cdot 7.5MgO\cdot Al_2O_3\cdot 7.5ZnO\cdot Al_2O_3\cdot 65SiO_2$-$AlPO_4$; up to approximately 70 wt% of the original SiO_2 content was substituted by $AlPO_4$. With an increasing amount of $AlPO_4$ the c lattice constant increases linearly, the CTE value increases slightly, and the a axis stays almost constant. The overall result of these investigations was that a broad field of compositions exists within the system Li_2O-MgO-ZnO-Al_2O_3-SiO_2-P_2O_5 for which glasses can be melted and transformed into glass ceramics with low CTE values. Therefore, the requirement for the CTE value may be met by many compositions, thus providing a relatively high degree of freedom for the optimization of the other requirements.

From the optimization cycles it became clear that some requirements were met rather easily, whereas others depended on the final composition chosen. The less critical requirements were those for Young's modulus, chemical resistance, and probably that for polishing. Young's modulus would vary only slightly within the predetermined composition field (its value would be around 90 GPa), therefore no attention was paid to this property during further development. Chemical resistance with respect to the proposed cleaning operations turned out to be sufficient in a rather broad composition field, so this property received lower priority. Preliminary polishing tests indicated that there were good chances that these glass ceramics could be finished to high optical figures. These results agreed with those published in the meantime about the glass ceramic Cer-Vit®, a material which had been developed for telescope mirror blanks by the competitor Owens-Illinois Co. and which was tested by different laboratories [4.8, 9]. The results were also confirmed by *Duke* and *Chase* [4.10] who obviously were engaged in a similar research project. The investigation of the requirement for good adhesion of an alu-

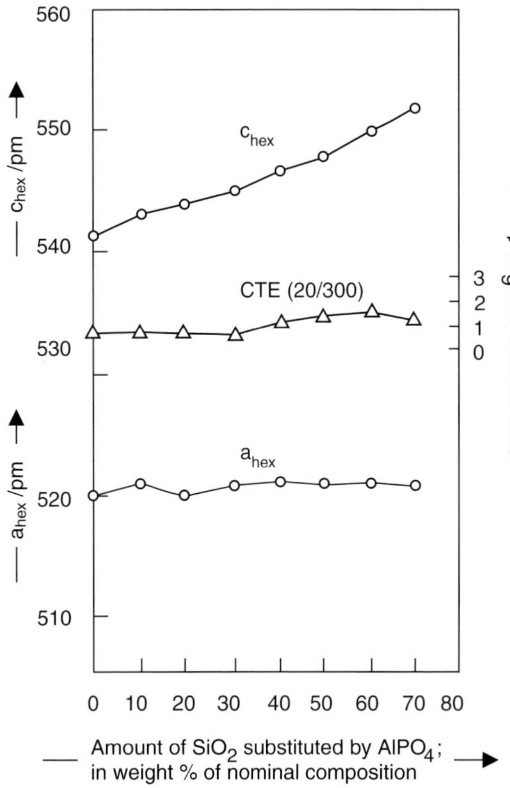

Fig. 4.1. Lattice constants and coefficient of thermal expansion, CTE (20/300), of $20Li_2Al_2O_4{\cdot}7.5MgAl_2O_4{\cdot}7.5ZnAl_2O_4{\cdot}65SiO_2$ in dependence on the partial substitution of SiO_2 by $AlPO_4$

minum coating to the polished substrate was postponed to a later stage; as aluminum adhered sufficiently well to SiO_2 glass and borosilicate glass it was hoped that this was true, too, for the new glass ceramic being also rich in SiO_2.

The selection of the final composition, therefore, was concentrated mainly on the optimization of two properties: transparency and homogeneity. *Tashiro* and *Tagaki* [4.2] had shown that transparency may be increased considerably by choosing a combination of ZrO_2 and TiO_2 as the nucleating agent in amounts of about 2 wt% each instead of 5 wt% TiO_2, which so far had been the preferred choice. The use of a combination of the two nucleating agents had the additional advantage that the transformation of the base glass into an h-quartz s.s. containing glass ceramic starts at a much lower temperature [4.1] compared with the same base glass composition nucleated by TiO_2 alone. This widens the time–temperature field in which h-quartz s.s. crystals form the main crystalline phase of the glass ceramic.

Further optimization of the transparency of the glass ceramic cannot be performed independently of the optimization of the homogeneous transformation of the base glass into the glass ceramic. General rules which are important to consider are:

(a) reduce the TiO_2 content to reduce coloration;
(b) choose a sufficiently high concentration of the nucleating agents ZrO_2 and TiO_2 so that the nucleus density is as high as possible, resulting in h-quartz s.s. crystals that are as small as possible, thereby reducing light-scattering effects;
(c) as the solubility of ZrO_2 in the base glass compositions is small, choose one which has a relatively high ZrO_2 solubility;
(d) try to match the refractive index of the residual glass with the mean refractive index of the crystalline phase because light scattering depends on the difference of both indices (the h-quartz s.s. crystals have hexagonal symmetry and are usually birefringent; make sure that birefringence is small).

The requirement for a base glass composition that is meltable with optical quality could not be addressed directly during the laboratory development project as small-scale melts with volumes in the range of liters always tend to be inhomogeneous. When glasses were melted in the laboratory their meltability was compared with that of optical glasses melted under the same conditions and for which it was known that high homogeneity is achieved in large melting units. The final decision, however, had to be postponed to trials in larger melting units.

The requirement for homogeneity concerns not only the question of whether a base glass can be melted with optical homogeneity, but also the question of whether a block of a homogeneous base glass can be transformed homogeneously into a glass ceramic; i.e., whether all volume elements of a base glass block having the same properties regarding the requirements for optical homogeneity will also meet these requirements after ceramization. This question can be investigated by laboratory experiments. As the h-quartz s.s. crystal containing glass ceramic represents a metastable phase assemblage that undergoes further transformations after prolonged heat treatment, the goal of the investigation must be to find a composition that can be transformed into an h-quartz s.s. glass ceramic with constant properties under widely varying ceramization conditions; the wider the time–temperature conditions may be varied without changing the specific glass ceramic properties the less stringent will be the requirements for temperature homogeneity during processing. The minimum target that had to be reached was to provide a glass ceramic that maintains its properties even under varying ceramization conditions, which correspond to the technical limit to realizing homogeneous temperature distributions in furnaces of 5 m in diameter and 3 m in height at heating and cooling rates of about 0.1–6 K/h. As it was not known at the time of the laboratory development how low the temperature inhomogeneities can be kept in such large furnaces and at temperatures up to about 900 °C, the goal was to find a composition as insensitive to temperature–time variations as possible; compositions allowing $\Delta T > 10\,\mathrm{K}$ certainly would be preferred.

After several cycles of optimization Petzoldt finally came up with a composition (in wt%) of: 57.2 SiO_2, 25.3 Al_2O_3, 6.5 P_2O_5, 3.4 Li_2O, 1.0 MgO, 1.4 ZnO, 0.2 Na_2O, 0.4 K_2O, 0.5 As_2O_3, 2.3 TiO_2, and 1.8 ZrO_2. This composition was the best choice out of a broad composition field which was patented [4.11]. The corresponding glass ceramic was called Zerodur® and some of its properties were described in [4.12]. Its key property, the expansion characteristic, is displayed in Fig. 4.2 using a more recent measurement. Assuming that all components that are suited for solid solution formation are incorporated into the h-quartz s.s. crystals, the crystalline phase would amount to 90 wt%, consisting (in wt%) of: 57.2 SiO_2, 20.5 Al_2O_3, 6.5 P_2O_5, 3.4 Li_2O, 1.0 MgO, and 1.4 ZnO; 0.2 Na_2O, 0.4 K_2O, 0.5 As_2O_3, and an excess amount of 4.8 Al_2O_3 would form the residual glass, while 2.3 TiO_2 and 1.8 ZrO_2 form the crystalline nucleating phase. However, the investigations of the glass ceramic obtained under the optimal ceramization conditions show that it contains only 70 vol% crystalline phase. The composition of this phase is deduced to be (in wt%) 59.4 SiO_2, 11.6 $AlPO_4$, 20 $LiAl_2O_4$, 4.7 $MgAl_2O_4$, 4.3 $ZnAl_2O_4$. A comparison of this composition with the data of the pseudobinary system in Fig. 4.1 shows that the compositions do not differ too much and that only limited use is made of the potential substitution (about 70 wt%) of SiO_2 by $AlPO_4$.

Figure 4.3 shows the transmission curve of Zerodur® at 5 mm and 25 mm thickness for a more recent production sample. The curve for the 5 mm thick sample is comparable to the one published in 1970 [4.12]. The transparency is sufficiently high to determine the residual strain after ceramization, even in rather thick samples. Thus, the transparency requirements are met by Zerodur®.

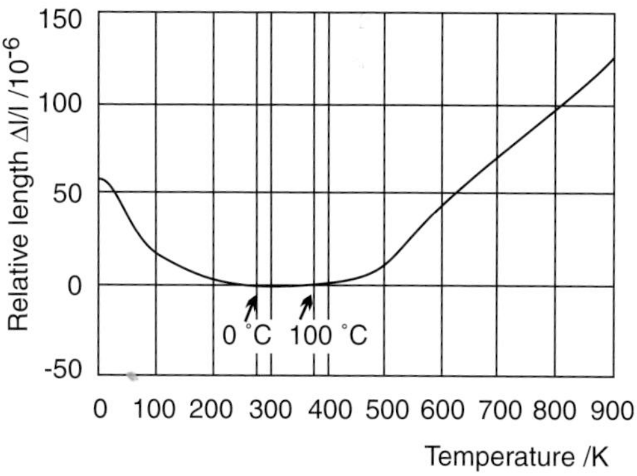

Fig. 4.2. Relative length variation of Zerodur® in dependence on the temperature

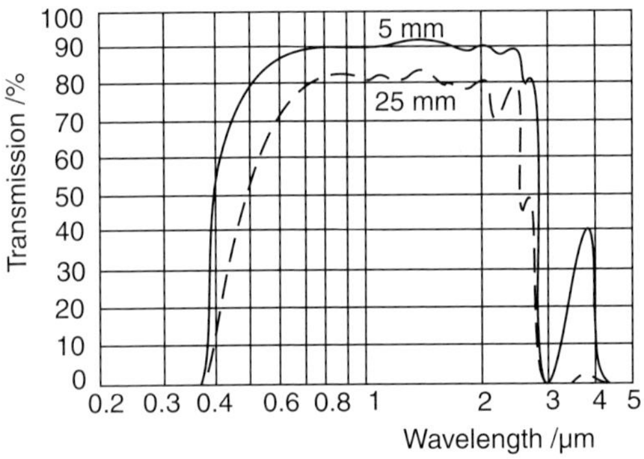

Fig. 4.3. Optical transmission of Zerodur® for two sample thicknesses

The insensitivity of Zerodur® to variations in the time–temperature history during cooling of the base glass and ceramization was demonstrated by the following experiments. One series of experiments determined the time–temperature conditions that characterize the transformations of the base glass to the h-quartz s.s. glass ceramic, on the one hand, and the h-quartz s.s. glass ceramic to the keatite s.s. glass ceramic, on the other hand. Figure 4.4 shows that there is a metastability region for the h-quartz s.s. glass ceramic that

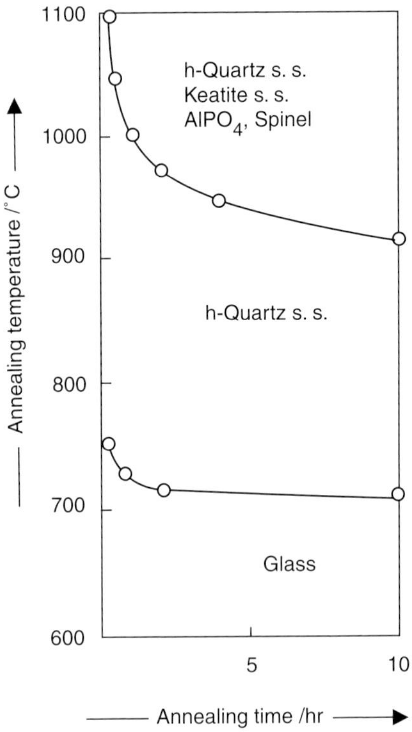

Fig. 4.4. Stability range of h-quartz s.s. containing glass ceramic in a time–temperature-transformation diagram for the Zerodur® composition

spans over 200 °C for constant times. This is a rather broad temperature interval suggesting that it should be possible to produce constant properties even for varying ceramization conditions.

In a second series, the variation of the CTE (−30/70) was investigated under time–temperature conditions considered to be close to the optimum for quick ceramizations. Figure 4.5 displays the CTE (−30/70) variation for three different temperatures and ceramization times between 2 h and 100 h (heating was from 600 °C to the nominal temperature with 8 K/h; samples taken out of the furnace at the end of the experiment). At 800 °C the variation of the ceramization time between 2 h and 100 h had no significant effect on the CTE (−30/70). Similarly, within ceramization times of 2–24 h, an increase of the temperature from 800 °C to 830 °C had no significant effect on the CTE. Thus, the time–temperature conditions can be varied to a considerable degree without effecting the final CTE value. Figure 4.5 also shows that choosing even larger variations of the ceramization conditions will start to change the CTE value significantly; but the conditions necessary to obtain these results are far from usual time–temperature inaccuracies in a well-controlled optical glass annealing furnace.

All these results indicated that a transformation of an optically homogeneous base glass into an h-quartz s.s. glass ceramic would yield a similarly homogeneous material even for large castings, which had to be ceramized at much lower temperatures over a period of approximately 5000 h.

With these data at hand, Schott informed the MPI that it was confident of realizing the proposed project successfully and that it would be interested in receiving the order for the fabrication of several mirror blanks.

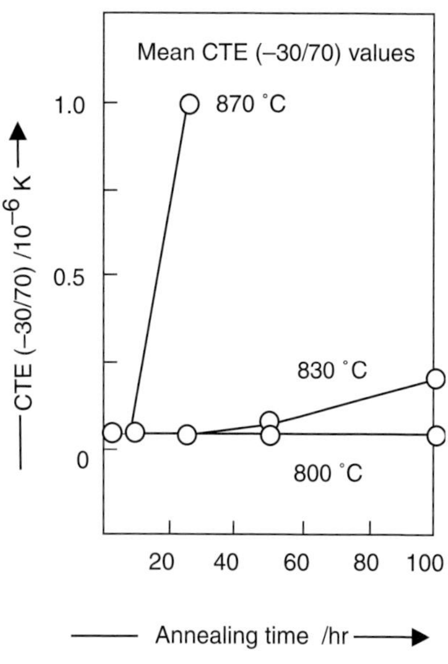

Fig. 4.5. Variation of the CTE (−30/70) of Zerodur® in dependence on the ceramization conditions; all samples heated with 8 K/h to indicated temperatures and quenched to room temperature at the end of the holding period

The main challenge of course still remained; would it be possible to produce castings with 4 m in diameter and almost 1 m in thickness?

4.1.2 Development of the Technology for Large Castings

While Zerodur® was being developed in the laboratory, the negotiations with the MPI continued. The potential order from the MPI would consist not of one large mirror blank with a diameter of 3.6 m but of a package of eleven mirror blanks altogether; as well as one mirror blank with a 3.6 m diameter the order would include two mirror blanks with 2.3 m diameters and eight smaller ones of different sizes. This was a fortunate situation as it allowed Schott to start the technology development with blanks of smaller sizes and to increase the sizes progressively.

The German Federal Ministry of Research and Technology became a third negotiating part, which substantially reduced the financial risks that the partners had to accept.

On the basis of the laboratory results, Schott accepted the order from the MPI in November 1968 and started to develop the technology for producing large castings. The first step consisted of planning and building a discontinuous tank, molds, the cooling and ceramization furnaces, handling and transportation equipment, machining equipment, and the logistics for the fabrication of monoliths of up to 2.5 m in diameter. Especially, the design of the melting unit required considerable alterations with respect to existing discontinuous tanks; the most important alterations necessary were changes in the refractories for the tank, the construction of the tank, the casting technology, and the mold materials.

The preparations for the first casting period were completed in February 1969 when the melting of the first batch started. A series of castings followed which was so successful that this development period ended in December 1969 with the conclusion that 2.5 m blanks can be produced with all properties lying within the required specifications.

This period, of course, was full of nervous and/or excited activities. The melting of the batch and the refining and homogenization of the melt were controlled in short intervals by several methods. The chemical composition of the glass was determined several times a day; this gave data about compositional changes due to the corrosion of the refractories as well as to selective evaporation. These data were supplemented by measurements of the viscosity and the glass transition point. The influence of chemical deviations from the specified composition on the properties of the glass ceramic were controlled by DTA and a fast ceramization program; the latter had been developed to provide within 24 h the CTE value, the crystal size, and the crystalline phase volume, which were expected for a proposed five to eight month ceramization program.

Numerous data, accumulated within a short period of time, were not only used to control the production parameters but they were at the same time analyzed for potential improvements. Some of these improvements were realized immediately during the melting period while others became evident only later when more and more results became available. Among the new data obtained were those from tests performed by Carl Zeiss on the polishing, the coating with aluminum, and the repeated cleaning of coated Zerodur® and the deposition of new aluminum coatings. Polishing was possible down to optical figures of $\lambda/20$; compared with borosilicate substrates the polishing time was reduced. The higher hardness of Zerodur® increased the time of polishing, but this increase was more than compensated for by the shorter time necessary for inspection. The adhesion of aluminum coatings was good. For the removal of the aluminum coating a cleaning procedure using NaOH and chromosulfuric acid was proposed; a test that comprised six cycles of coating and cleaning did not show any detectable change in the optical figure. Later a procedure was developed based on the two main ingredients NaOH and HCl and, thus, is better suited for technical processes; see Sect. 4.5.3. By this procedure aluminum coatings can be replaced 100 times without affecting the optical figure.

The homogeneity measurements of the 2.5 m castings attracted particular attention. Visual inspection of the base glass blanks with polarized light indicated good homogeneity, so they were processed for ceramization. During ceramization the temperature of the furnace was kept constant within 3.5 K over the furnace volume. This was achieved by using heating rates as low as 0.1 K/h and taking appropriate account of the heat of 245 kJ/kg which is generated during ceramization by the phase transformations. After ceramization, samples for property measurements were taken from plates on the top and the bottom of the blank, from the side walls, and from the central holes which were machined into the castings after ceramization. Measurements of the refractive index for samples taken from the top and bottom plates of a very good casting indicated variations of only $\pm 15 \times 10^{-6}$; at that time these values were considered to be very good ones, while nowadays values of a few 10^{-6} are achieved regularly. Measurements of the CTE (−30/70) were performed with a calibrated push rod dilatometer made of fused silica; 33 samples taken from the locations mentioned previously yielded a mean value of $+0.015 \times 10^{-6}\,\mathrm{K}^{-1}$ with all but three values lying within the accuracy range of $+0.015 \pm 0.025 \times 10^{-6}\,\mathrm{K}^{-1}$.

The mean value of $+0.015 \times 10^{-6}\,\mathrm{K}^{-1}$ was considered to be excellent; it is almost a factor of 30 smaller than the value for fused silica, the material with the best CTE value so far. At least of similar importance was the fact that the mean value was better by a factor of 10 than the specification of $\pm 0.15 \times 10^{-6}\,\mathrm{K}^{-1}$. If one succeeded in reproducing this value for the 3.6 m blank, the great expectations concerning the performance of the new telescope could even be surpassed. When all the data mentioned before had been collected it became obvious that the new glass ceramic Zerodur® was well suited to shift the resolution of state-of-the-art telescopes far beyond the limitations of the time. Only one question remained: would it be possible to scale the production process from 2.5 m blanks to 4 m blanks without sacrificing one or more of the excellent properties?

While the glasses and glass ceramics produced during the first development period were being processed, and data collected and analyzed, the preparation of the second development period had started. Larger production facilities had to be designed, planned, constructed, and put into operation. The construction of the new discontinuous tank, which was considered to be particularly critical, was given special attention. The old tank had a melting capacity of 12 t of glass; the new one should have one of 48 t. Consequently, the construction of the tank had to accommodate this increase by a factor of four. Changes in the construction were followed by many others; these included, to mention only a few, new refractories, adjustment of the melting, refining, and homogenization procedures, a new batch composition to make provisions for the altered influences of corrosion and selective evaporation on the final glass composition, new mold designs, and new casting technology to keep the pouring time within several hours. While many parameters could be tested either by small-scale experiments or isolated tests of single steps, the effectiveness of the newly designed, overall production process could only be tested in a full-scale experiment. Though this is a normal approach in every development, the enthusiasm for testing many variables broke down quickly at costs of about DM 300 000 (in 1971) for each experiment. Therefore, the pressure on the development team was very high to find the good production parameters – which the team hoped existed – within a few trial runs. Otherwise, the total project budget would soon have been exceeded.

The first production test was performed in the autumn of 1971, but it failed due to deficiencies in the pouring technique. Further testing showed that the construction of the tank had to be changed. These changes had been performed by February 1972 and a testing series could start in March 1972; it included two castings of which the first one was attended by guests and public representatives. The technical procedures went very well in both cases, whereas inspection of the castings indicated that their homogeneity was not sufficient. Thus, any further tests were stopped. It was concluded that the construction of the tank as well as the melting procedure needed considerable improvement to achieve the required homogeneity.

The observed inhomogeneities were analyzed and their reasons discussed. Ideas for improvements were proposed, tested, and finally implemented, resulting in a new tank, which was ready for production at the end of 1973. A three-run series of tests was performed and when the castings could be inspected it became clear that it had been successful. From what was known at this point in time it was concluded that the homogeneity of the glass blanks was so good that the specifications would be met after ceramization. Later this turned out to be true.

In the following, the production of the mirror blank cast on December 28, 1973, which later became the mirror blank of the 3.6 m mirror of the MPI telescope on top of Calar Alto, Spain, is described in more detail. For the description the present tense is used as the main process steps have not changed over the years and are still valid; the numbers given for process parameters are those used in 1973/74; they may have changed in the meantime. Melting of the batch occurs at about 1550 °C, for fining the temperature is raised slightly to about 1600 °C, and during homogenization the temperature is lowered to about 1400 °C; the whole process takes about three weeks during which the composition is controlled by the above-described means. The next step is the casting process. The hot glass melt is poured into a well-isolated mold which has been heated up to about 1200 °C. During the pouring period, which takes several hours, the mold is filled with approximately 27 t of hot glass. At the interface of the mold material and the glass melt the temperature is approximately 1300 °C; here crystals are nucleated and grow, but, fortunately, the growth velocity is rather small, so that the thickness of the layer formed during pouring is small. Figure 4.6 shows the tank and the pouring process of the casting performed on December 28, 1973.

When the mold is filled with the stipulated glass volume, the pouring process is stopped and the mold is moved out of the area under the tank. Then the process of rapid but semi-controlled cooling to about 900°C starts. This is performed first by raising the mold cover and later by using hoods of different geometries. The process has to be performed as rapidly as possible to keep the crystalline layer which grows during cooling as thin as possible. Especially, the temperature interval around 1100 °C has to be passed quickly as crystal growth velocities have their maximum at this temperature. So, all crystals that had formed at the interfaces during pouring will grow at much higher velocities and a crystalline layer which has a much smaller CTE value than the bulk glass will form. On the other hand, the cooling has to be controlled, so that no local reheating of interior volume elements of the undercooled melt occurs. In particular, no reheating of a volume element should occur once its temperature has been below approximately 850 °C as then internal nucleation and crystal growth are initiated and the ceramization process starts in an uncontrolled manner.

Rapid cooling to about 900 °C is completed after several hours. At this temperature, crystal growth velocities are three orders of magnitude smaller than at 1100 °C and there is no further need to hurry. The very viscous undercooled melt is now prepared for controlled cooling to room temperature in a cooling furnace. This mainly means that the side walls of the mold are removed, so that they will not shrink onto the glass during cooling and destroy the block. After removal of the side walls the block is moved into the cooling furnace where it is cooled to room temperature in 4.5 months.

After cooling, first of all the crystalline layer on the surface of the block as well as cracks initiated by this layer and running into the block are eliminated

Fig. 4.6. Casting of a 4-m blank on December 28, 1973

to avoid any further subcritical crack growth. Then the quality of the block is inspected to find out whether it is worth processing it any further into a mirror blank. If this is the case, further machining is performed to prepare the block for ceramization. By all these measures the weight of a block of 27 t is reduced to about 18 t. This 18 t block is ceramized over a period of eight to nine months. During the ceramization the 4 m blank shrinks in its diameter by 40 mm, changes its CTE from $4.0 \times 10^{-6}\,K^{-1}$ to almost zero, and releases 4 400 000 kJ within 40 days.

During the ceramization of the casting the temperature in the ceramization furnace is controlled by 31 thermocouples; the largest difference measured between any two of them during ceramization is 2.4 K. Thus, the changes that have been made since the ceramization of the 2.5 m blanks improved the temperature homogeneity in the furnace substantially. In June 1975, the ceramization of the 4 m blank was completed; when the furnace was opened the team saw with much relief that the block looked undamaged.

The first impression that the ceramized block was of good quality was confirmed by the subsequent measurements. So the machining of the final approximate shape was started. This machining consists mainly of milling the approximate contour of the spherical mirror with radius 24.5 m into the flat disk and of drilling the central hole. As these machining operations change the low strain field of the flat disk considerably, local strains build up that exceed the specifications; therefore, the rough machined spherical mirror has to be annealed once more. This operation lasts almost another three months. Further machining follows until the mirror blank has its stipulated geometry with a surplus of 1 mm. All the machining operations reduce the 18-t glass block that is used for ceramization to 14.1 t.

The mirror blank, which was then transferred to Carl Zeiss for optical figuring, had a diameter of 3604 mm and a thickness of 593 mm.

Measurement of the CTE (−30/70) from 60 locations chosen in a similar way as for the 2.5 m mirror blank again gave a mean value of $+0.015 \pm 0.025 \times 10^{-6}\,K^{-1}$; this time not one value lay outside the accuracy range.

Late in 1975 the project of manufacturing a 3.6-m mirror blank from the new optical glass ceramic Zerodur® was completed, about nine years after the start of the materials development in the laboratory and about seven years after the start of the development of the production technology. Since the start of the production technology, development costs of about DM 10 000 000 had accumulated; funds of DM 3 800 000 by the German Federal Ministry of Research and Technology supported the development and helped decision-making considerably at various stages. About 150 people were involved in the project. The precision machining at Carl Zeiss took another four years and it was in 1984 that the telescope was installed on the top of the 2160 m high Calar Alto in the southeast of Spain. When the first observations were available the scientists noticed with great relief as well as with excitement that the new telescope met all their hopes and expectations.

At the end of the project, Schott had created a new material for precision optics applications, had sold some telescope mirror blanks, but had also quite an amount of unsold material of different quality in stock. It was now up to the sales people to find other applications for Zerodur® than telescope mirror substrates, thereby creating new markets for this new material. Relatively soon two new applications were found which will be described in more detail in Sect. 4.6; one was mirror substrates for reflective optics in microlithography, for which castings of approximately 500 mm in diameter were needed, and the other was housings for laser gyroscopes, for which continuous bars of cross sections of about $140 \times 40\,mm^2$ were the best delivery geometry. Schott soon ran out of stock for the geometries and quality required, and a new melting period had to be planned and started. As the sizes required were small, a continuous process in an optical tank was considered to be the most economic production technology. The first continuous production run was started in 1981 and severe difficulties were encountered with respect to homogeneity and – totally new compared with the discontinuous process – bubbles. Analyzing the melting process and changing parameters in a longstanding and tough process, the required quality was finally obtained. It took a few more melting periods until the continuous melting process was so well understood that production of Zerodur® base glass became a routine operation.

4.1.3 Properties

Those customers who used Zerodur® in the above-mentioned two applications, but also others who wanted to use Zerodur® in applications with minor sales volume, often asked questions about properties that so far had not been measured. In response to these requests Zerodur® was characterized with time by more and more methods as well as by different groups. The most important results are outlined in the following; additional information is given in the references [4.13, 14].

Thermal Properties

The key property of Zerodur® is of course its thermal expansion characteristic, which is very low in the temperature interval -50 to $+100\,^{\circ}$C, see Fig. 4.2. Zerodur® is nowadays usually characterized by its CTE (0/50) which serves to define three expansion classes. Originally, the three classes were given by the CTE (0/50) value ranges of $0 \pm 0.05 \times 10^{-6}\,\mathrm{K}^{-1}$, $0 \pm 0.10 \times 10^{-6}\,\mathrm{K}^{-1}$, and $0 \pm 0.15 \times 10^{-6}\,\mathrm{K}^{-1}$ for classes 1, 2, and 3, respectively. With increasing ceramization know-how Schott was able always to guarantee class 2 material, so that class 3 was not of interest anymore. On the other hand, Schott could accept specifications with a CTE (0/50) value range of $0 \pm 0.02 \times 10^{-6}\,\mathrm{K}^{-1}$ without driving prices too high; this range then became class 0, so that nowadays the three expansion classes 0–2 are offered.

The larger the single pieces become in precision optics applications, the more concern is expressed with regard to their homogeneity. Due to the high transparency of Zerodur®, inclusions, bubbles, and striae are easily detected so that only soft long-range inhomogeneities are difficult to detect. These potential inhomogeneities are checked by taking samples for CTE (0/50) measurements from different locations and evaluating the scatter in the values obtained. The determination of the number and locations of the samples taken is part of the agreement between the customer and Schott. In general terms, the following specifications are offered when the size of a piece is expressed in weight (the density of Zerodur® is 2.53×10^3 kg/m^3): CTE (0/50) variations $\leq 0.03 \times 10^{-6}\,\mathrm{K}^{-1}$ for pieces up to 18 t, $\leq 0.02 \times 10^{-6}\,\mathrm{K}^{-1}$ up to 6 t, and $\leq 0.01 \times 10^{-6}\,\mathrm{K}^{-1}$ up to 0.3 t. The accuracy of the CTE (0/50) measurement procedures is $\pm 0.01 \times 10^{-6}\,\mathrm{K}^{-1}$.

Other thermal properties of interest are, e.g., thermal conductivity and specific heat; their values at 20 °C are 1.46 W/m K) and 0.80 J/g K, respectively. The reader interested in the temperature dependence of these properties is referred to [4.13].

Zerodur® has been designed for applications in the temperature interval from −50 °C to +100 °C. However, during processing at the customer's shop or for some applications, for example space applications, Zerodur® is also handled outside of this main temperature interval. Except for high precision components, such handling is possible for long periods of time up to 600 °C. If an exposure to even higher temperatures is required, Zerodur® may be subjected to such temperatures for well-determined periods of time without loosing its properties. This, however, must not last indefinitely long as the transformation of the glass into a glass ceramic, which has been stopped at a well-defined point, will proceed. Figure 4.7 illustrates how the expansion curve of Zerodur® changes under additional heat treatments at 800 °C. In case of the fabrication of high precision components special care has to be taken with respect to two temperature intervals, as will be explained in more detail later in this section and in Sect. 4.5.1.

Mechanical Properties

When compared with normal glasses, Zerodur® can be classified as a "strong glass"; its Young's modulus with 90.3 GPa and its Knoop hardness of 620 belong to the higher values known for glasses. Its density is 2.53×10^3 kg/m^3; its Poisson's ratio is 0.24. As for glasses, statements about the strength of Zerodur® are difficult to make as strength is not a material property but depends on parameters such as surface flaws, the area tested, environmental conditions, and stress rate. For most applications a design strength of 10 MPa

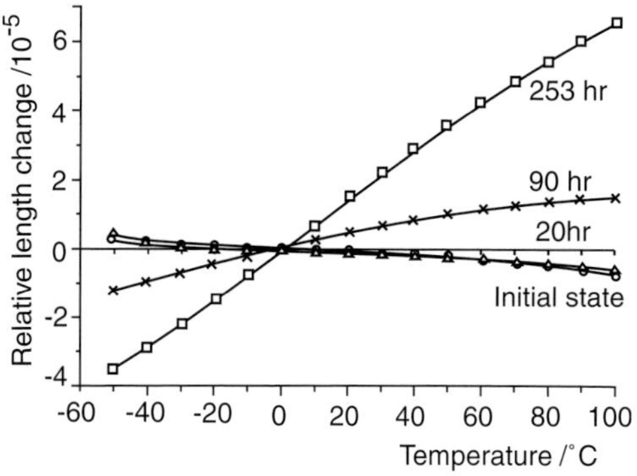

Fig. 4.7. Rotation of the $\Delta l/l$ curve of Zerodur® by additional isothermal heat treatments at 800 °C for three holding times

is appropriate; under special circumstances, however, lower values have to be taken. Also, when higher design values are desired, a special evaluation of the load conditions may indicate whether, and under what conditions, higher design strengths are acceptable. The parameters of the stress corrosion exponent and the fracture toughness, which are important to such an evaluation, have been determined to be 51.7 and 0.85 MPa $\mathrm{m}^{1/2}$, respectively, in air [4.15]. Further parameters such as the Weibull modulus and the characteristic value, which characterize certain surface conditions, are given in [4.15] and [4.16]. For a more detailed evaluation of the strength of Zerodur®, [4.17, 18] may be helpful.

Thermomechanical Properties

For large mirror applications the response of materials to combined thermal and mechanical loads is often evaluated, and materials are positioned in an evaluation diagram according to their ability to maintain or to regain their optical figure with respect to such combined loads. Differences in the ease to improve the constancy of the optical figure by constructive measures, as, e.g., light weighting, forced cooling, or active support, which are due to different fabrication techniques, are not taken into account.

Ayer [4.19] defined two now widely accepted figures of merit (FOM) to characterize the sensitivity of different materials with respect to such loads; one FOM serves the case of transient situations, the other serves that of steady state conditions. The two FOMs are defined as

$$\text{transient:} \quad \mathrm{FOM_t} = \rho \frac{(1-\mu^2)}{E}\, \alpha\, \frac{C_p \rho}{\lambda} \qquad \text{in } 10^{-9}\,\frac{\mathrm{s}^3}{\mathrm{m^4 K}}\,,$$

$$\text{steady state:} \quad \mathrm{FOM_s} = \rho \frac{(1-\mu^2)}{E}\, \frac{\alpha}{\lambda} \qquad \text{in } 10^{-15}\,\frac{\mathrm{s}^2}{\mathrm{W/m}}\,,$$

where ρ is the density in $10^3\,\mathrm{kg/m^3}$, μ is Poisson's ratio; E is Young's modulus in GPa, α is the CTE in $10^{-6}\,\mathrm{K}^{-1}$, C_p is the specific heat in J/(kg K), λ is the thermal conductivity in W/mK. The higher the FOMs of a material the more sensitive it is to thermomechanical loads. If α is taken as $0.05 \times 10^{-6}\,\mathrm{K}^{-1}$, the values of $\mathrm{FOM_t}$ and $\mathrm{FOM_s}$ for Zerodur® are $1.6 \times 10^{-9}\,\mathrm{s^3/m^4K}$ and $0.8 \times 10^{-15}\,\mathrm{s^2/W/m}$, respectively. These values are among the best values obtained for potential mirror substrate materials. Further, it should be noted that these FOMs are very sensitive to variations in the CTE values for all low-expansion materials with CTE values close to zero. If Zerodur® of expansion class 0 is taken, the FOMs improve at least by a factor of 2.5.

Optical Properties

As already mentioned the most important optical property of Zerodur® is its transmission; see Fig. 4.3. Low absorption coefficients in a wide wavelength range allow easy control of the internal homogeneity as well as of the presence of inclusions even in large pieces. The absorption at the short wavelength side of the visual transmission range is mainly due to the absorption by charge transfer processes between iron impurities and titanium; a smaller contribution results from scattering processes at the interface between the crystals and the residual glass. The Rayleigh scattering under 90°, R90, at 404.7 nm is, for example, R90 $= 23 \times 10^{-3}\,\mathrm{cm}^{-1}$. Some values of optical constants are: refractive index $n_\mathrm{d} = 1.5424$, Abbe number $\nu_\mathrm{d} = 56.1$, dispersion $n_\mathrm{F} - n_\mathrm{C} = 0.00967$, and stress optical coefficient K at a wavelength of 589.3 nm $K = 3.0 \times 10^{-6}$/MPa. Requirements for high optical homogeneity levels are accepted on special requests. Pieces with a refractive index homogeneity level H4, corresponding to variations in the range $\pm 1 \times 10^{-6}$, can be supplied.

Chemical Resistance

Chemical resistance is tested according to standard testing procedures for optical glasses. In cases where international standards have been established, testing is performed according to these standardized testing procedures, in all other cases the procedures described in the *Schott Catalogue of Optical Glasses* [4.20] are used. The chemical resistance of Zerodur® is good as is demonstrated by the following data for the three most important tests: hydrolytic resistance (ISO 719) – class 1; acid resistance (ISO 8424) – class 1.0; alkali resistance – class 1.0.

As mentioned before, mirror coatings have to be replaced regularly for some applications. In these cases the coatings are dissolved chemically. The proposed procedure is described in more detail in Sect. 4.5.3.

Electrical Properties

Compared with technical glasses the electrical resistivity of Zerodur® is rather poor. The temperature for a resistivity of $10^8\,\Omega\,\mathrm{cm}$ is 178 °C. The dielectric constant ϵ and the loss factor $\tan\delta$ at 1 MHz are 7.4 and 0.015, respectively. These data are not surprising. β-eucryptite, a crystalline material with the special h-quartz s.s. composition $Li_2O{\cdot}Al_2O_3{\cdot}2SiO_2$, is known to be a good one-dimensional ion conductor [4.21], and due to the structure of h-quartz also other solid solution compositions are expected to behave similarly. So, neither the residual glass phase nor the crystalline phase are expected to block ionic conductivity effectively.

Helium Permeability

An important application for Zerodur® is laser gyroscopes as will be described in Sect. 4.6. In this case Zerodur® forms the cavity of an He–Ne laser; for its operation it is important that the concentration of the He gas is kept constant over the lifetime of ten years of the gyroscope. On the other hand, He has the ability to diffuse rather easily through materials. Therefore, the He permeability of Zerodur® and other competing materials has been investigated. Figure 4.8 demonstrates that the He permeability of Zerodur® is more than two orders of magnitude lower than that of the other low-expansion materials ULE® (7971) and fused silica (7940).

4.1.4 Internal Quality

The internal quality of Zerodur® is determined by four factors: inclusions, striae, bulk stresses, and long-range homogeneity. Detailed data on those properties are given in [4.13].

The term "inclusion" comprises particles, i.e., stones or crystals, and bubbles. In general, the inclusion level of Zerodur® is very low. In most applications small inclusions do not have any effect on specified properties; they are, therefore, considered only when they are larger than 0.1 mm in diameter. Also larger, isolated inclusions normally do not effect bulk properties unless they lie very close to a functional surface. Inclusions with sizes larger than 0.1 mm in diameter are evaluated according to the following criteria:

- projected area of all inclusions in mm^2 per $100\,cm^3$,
- average number of inclusions per $100\,cm^3$,
- maximum diameter of individual inclusions in mm.

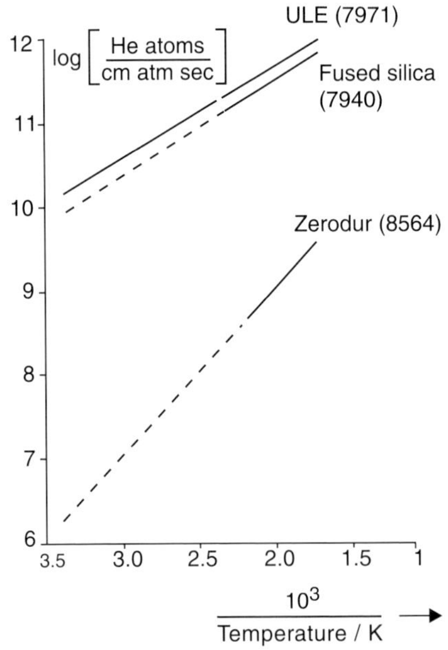

Fig. 4.8. Helium permeability of Zerodur® in comparison with two low-expansion glasses [4.22]

The third criterion is subdivided into different groups depending on the size of the Zerodur® part considered and on the location of the inclusion in that part; an inclusion found in an uncritical volume element may be larger than one found close to a functional surface and thus in a critical volume element.

With respect to these criteria a "standard" and a "special" grade are defined for normal applications. For applications with more demanding requirements, "inclusion classes" 3–0 have been introduced; they are identical with those defined for optical glasses [4.20].

Striae are transparent, locally limited regions, which deviate in their composition from that of the surrounding material. This composition inhomogeneity results in an inhomogeneity of the optical properties due to differences in the optical constants as well as to local stresses.

The stress-induced birefringence is used to characterize striae; five different quality levels are distinguished, depending on the optical path difference caused by the striae. The path difference is measured with white light and given in nm. The standard quality may have striae causing a maximum optical path difference of less than 60 nm. For applications with very strict striae requirements the stress-optical method is not sensitive enough; in these cases a shadow method is used, which is very sensitive to detecting even the faintest striae.

Bulk stresses are formed during the cooling of a glass ceramic block through its glass transition temperature region, which exists not only for glasses but also for glass ceramics with a residual glass phase. To achieve bulk stresses that are low and symmetrically distributed, Zerodur® articles are subjected to precision optical annealing. To characterize the different cooling histories of bulk and surface volume elements, the birefringence caused by the bulk stresses is measured at an edge distance of 5% of a characteristic length of the article under consideration and divided by the article thickness. Two levels are distinguished, each of which is subdivided into groups depending on the size of the article; stress birefringence values for these quality levels lie in the range of 15–4 nm/cm.

4.1.5 Delivery Shapes, Dimensions, and Tolerances

Zerodur® can be supplied in the form of disks, rectangular blocks, prisms, rods, and cut pieces. Thick disks that are self-supporting can be produced with up to more than 4 m in diameter and with volumes of about 8 m^3 (equivalent to 20 metric tons). Thin menisci that are not self-supporting can also be produced up to masses of about 20 t; the largest sizes produced so far were the 8.2-m menisci for the Very Large Telescope of the European Southern Observatory; see Sect. 4.3.1.

All basic geometric shapes can be obtained by precision machining using modern CNC processing equipment and high-performance diamond tools.

These shapes also include certain light-weight structures which combine the requirements of high stiffness with low weight and low heat capacity.

Zerodur® parts are machined to final shapes with tolerances in the range of mm. These tolerances are plus tolerances and their upper values depend on the size of the part; the following values, for example, are guaranteed for a disk with 1 m in diameter, a tolerance of 3.2 mm, a flatness of 0.6 mm, and a spherical profile of 1.2 mm. The surfaces of the parts are ground with diamond tools having grit sizes of D252 or D151. Parts machined within these tolerances are then usually shipped to an optical shop for optical figuring.

In some cases, tighter tolerances are required by optical figuring shops. Zerodur® thin menisci or plano-spherical blanks can also be supplied in near-net shape. Net shaping means linear dimension tolerances of 0.2 mm or smaller. The spherical grinding of such net-shaped blanks leads to profile tolerance zones ranging from 0.05 to 0.40 mm P-V. These tolerances are achieved not only for blanks of 1.0 m in diameter but also for thin menisci of 8.2 m in diameter.

4.1.6 Development of Zerodur M®

During the development of Zerodur® for the MPI telescope mirror blank the question was raised whether Zerodur® shows any effect of thermal ageing, which is common to many glasses. *Petzoldt* [4.23] performed some investigations but did not observe any effect. He concluded that one reason for that could be the relatively small amount of 30 vol% of the residual glass phase. But when Zerodur® was tested for other applications, a sensitivity of the length of Zerodur® to thermal history was observed.

Bennett [4.24] was the first to observe such an effect during thermal cycling between 20 and 300 °C. When a Zerodur® sample was cycled twice between the two temperatures, a length change was observed during the first cycle, while no further change was observed during the next cycle. Similar results were reported by *Gorski* [4.25] for thermal cycling experiments between room temperature and 500 °C: $\Delta l/l$ curves for heating and cooling coincide at temperatures above 250 °C but deviate from each other below 250 °C.

These results stimulated an independent investigation at Schott. First of all it seemed to be necessary to develop a new push rod dilatometer with a much higher sensitivity than that made from fused silica. Such a dilatometer seemed to be indispensable for two reasons. Firstly, the interferometric measurements used by *Bennett* [4.24] and *Gorski* [4.25] are time-consuming and expensive; they are not suited for quick and numerous experiments. Second, the effects observed are rather small; a good signal-to-noise ratio is essential to understand the phenomena and for future materials development. *Lindig* [4.26] built a new push rod dilatometer made of titanium silicate glass (ULE® 7971) with a sensitivity of 5 nm (for further details see [4.27]). Then the length variations of Zerodur® and experimental melts were investigated depending on a large variety of thermal histories. It was soon realized that the effects observed by Bennet and Gorski can be circumvented by choosing an MgO-free composition.

But for some time no MgO-free glass ceramic composition was found that had a similar expansion characteristic as Zerodur® and would thus have tentatively been able to substitute it. In the meantime, additional investigations have been published [4.28], the most thorough one probably being that of *Jacobs* et al. [4.29]. Unfortunately, the authors chose a presentation that is not suited to illustrate the phenomena but rather to confuse them. The main complaint concerns the fact that the authors switch permanently from constant heating/cooling rates to isothermal holdings, use the temperature axis as the time axis (a three-dimensional presentation would have been appropriate), and do not document properly when these switches occur. The most important observations of *Jacobs* et al. [4.29] are: (a) isothermal length changes are observed in the temperature interval from 130 to 207 °C, thus showing which temperatures probably were important to the observations of *Bennett* [4.24] and *Gorski* [4.25]; (b) there is a second temperature interval from −80 to +27 °C where similar effects can be observed. The temperatures of 207 °C for the upper temperature interval as well as −80 °C for the lower one need not necessarily be the temperature boundaries of the observed effects but resulted from the temperature limitations of the experiments.

While these new papers were published Schott still worked on the development of a Zerodur® variant free of the hysteresis effect in the upper temperature interval. Finally, a composition field was found and Schott presented its point of view of the hysteresis phenomena, *Lindig* and *Pannhorst* [4.27, 30]; by choosing different measurement procedures and presentations it was believed that a more consistent picture of the relevant effects was obtained than had been done by *Jacobs* et al. [4.29].

The most important results for users of Zerodur® are presented in Fig. 4.9. It shows that the relative length variation in the temperature interval from 300 °C to room temperature depends on the cooling rate. Particularly important is the observation that the final lengths at room temperature are different. Therefore, uncontrolled cooling from 300 °C to room temperature will result in different contraction/expansion behavior of single-volume elements and, therefore, will destroy previously finished optical figures. The change in optical figure can be avoided by controlled cooling. As was already indicated by the observations of *Jacobs* et al. [4.29], controlled cooling is not necessary in the whole interval from 300 °C to room temperature, which is investigated here, but only within the range from 300 °C to 130 °C. *Lindig* and *Pannhorst* [4.27] confirmed the lower temperature boundary observed by *Jacobs* et al. [4.29] and showed that the upper boundary of the temperature interval is 320 °C. Furthermore, it is important to know that once an optical figure has been obtained for a component with a well-defined cooling rate in the temperature interval from 320 °C to room temperature, this optical figure can always be re-established by reheating the component to 320 °C

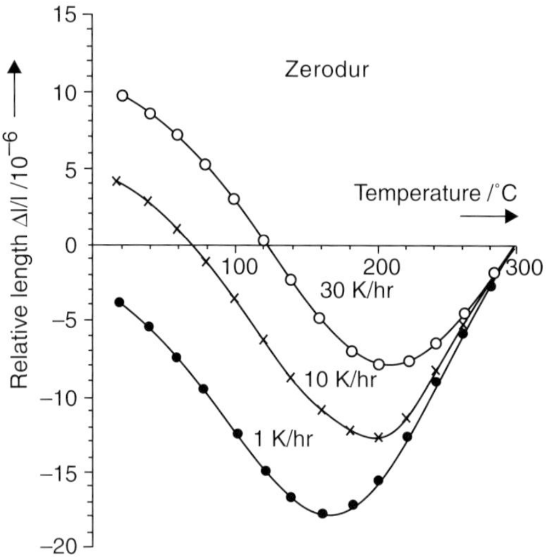

Fig. 4.9. Relative length variation of Zerodur® between 300 °C and 20 °C for three cooling rates

and cooling it to room temperature (or, at least, to 130 °C) with the original cooling rate. So, changes of the optical figure resulting from uncontrolled heat treatments during intermediate processing steps are not permanent but the initial optical figure can be restored by fine annealing from 320 °C with the original cooling rate.

There are, of course, applications for which processing steps have to be performed in the temperature interval between 130 °C and 320 °C and for which a final fine annealing from 320 °C is not possible, for example when coatings are not stable at this temperature. For such applications, which at the same time require very precise optical figures, a new Zerodur® variant, Zerodur M®, has been developed [4.31]. Its composition is free of MgO and does not show any hysteresis effect between 320 °C and room temperature within the accuracy of current measurement techniques (see Fig. 4.10) [4.32, 33]. The expansion characteristic of Zerodur M® is comparable to that of Zerodur®, (see Fig. 4.11), thus also guaranteeing very stable optical figures in precision optical applications.

A pilot production run was performed in 1985 and castings of high quality with sizes up to about 1 m in diameter were obtained. Although the property requirements are met for Zerodur M®, the CTE value is much more sensitive to variations in the time–temperature program during ceramization than is the case with Zerodur®. Further experience has to be gained with different loads and sizes of ceramization furnaces until Zerodur M® of expansion class 0–2 quality is obtained with the same confidence level as for Zerodur®. Most properties of Zerodur M® deviate only slightly, for most applications insignificantly, from those of Zerodur®. Larger changes are observed for the Knoop hardness (540) and in chemical resistance (acid resistance: class 4, alkali resistance: class 2). For further details the reader is referred to [4.14].

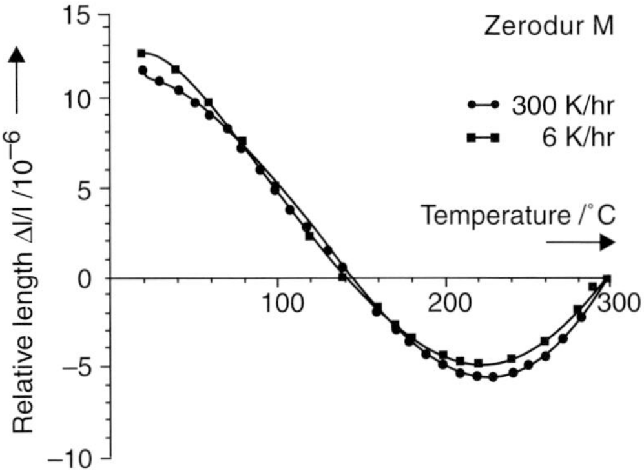

Fig. 4.10. Relative length variation of Zerodur M® between 300 °C and 20 °C for two cooling rates

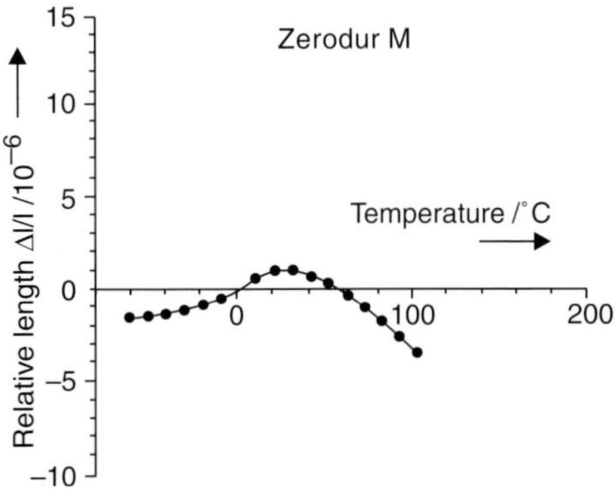

Fig. 4.11. Relative length variation of Zerodur M® in the temperature interval between −60 °C and 100 °C

Much less attention has been paid to the characterization of the hysteresis phenomena in the lower temperature interval [4.27, 29, 33–35]. Both glass ceramics, Zerodur® and Zerodur M®, show such phenomena of about equal size; the temperature interval is somewhat smaller for Zerodur M® (−80 °C to 10 °C) than for Zerodur® (−80 °C to 40 °C). Up to now it is not known whether these phenomena can also be avoided by a variation of the chemical composition as has been the case for the hysteresis phenomena in the upper temperature interval (130–320 °C).

A more detailed presentation of the hysteresis phenomena is given in Sect. 4.5.1.

4.2 Conventional Production of Zerodur®

Rüdiger Hentschel, Hartmut Höness, Rudolf Müller, Norbert Reisert

The conventional production of Zerodur® is characterized by the following main production steps. Generally, after the glass has been melted continuously or discontinuously in a melting tank, the hot melted glass is formed (into rods, blocks, or round disks) and then cooled down to room temperature in the glassy condition. By a subsequent thermal treatment, the transition is made from the glassy to the partially crystallized condition (ceramizing). In general, the final geometry is established with the help of cutting techniques, such as sawing, grinding, boring, etc. If it is necessary to establish a defined stress (compressive stress at the surface), the workpiece is subjected to a fine annealing process. Manufacturing controls and quality tests take place between the individual manufacturing steps.

The manufacturing steps described in the following are also the basis of the manufacture of Zerodur® blanks with special geometries (see Sect. 4.3).

4.2.1 Melting

The casting and hot-forming times of Zerodur® blanks may not be chosen freely (for details on this see Sect. 4.3). For this reason, the glass can be melted continuously with a relatively low throughput only for the manufacturing of small-volume pieces. In contrast, the throughput must be increased when large-volume parts are cast so that the casting times are not too long. The glass must, therefore, be removed almost completely from the melting tank when it is cast, which means that the melting of the glass can take place only in a discontinuously operated melting unit.

Continuous Melting

A continuously operated melting unit (see Fig. 4.12) consists of the following main components: melting tank with peripherals (insertion system, arch heating, side/bottom heating), refining chamber, stirring crucible, feeder, various heating and cooling systems, instrumentation and control with process computer.

The batch is placed into the melting tank (melt volume up to approximately 900 l) and melted using the arch heating and the electrical side/bottom heating (glass temperature approximately 1550 °C). The quantity of the mixture inserted corresponds to the quantity of glass removed (up to 1000 g/min) so that the glass level remains constant.

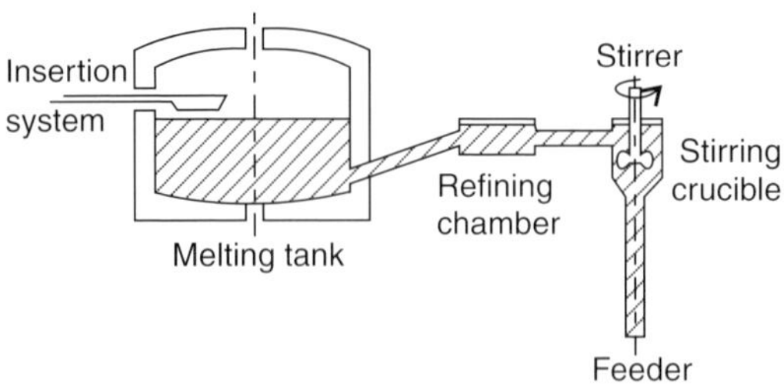

Fig. 4.12. Schematic representation of the structure of a continuously operated melting unit

Due to the removal of the glass, the melted mixture rises in the refining chamber. Raising the temperature to approximately 1600 °C causes the arsenic oxide to decompose, which produces the so-called refining bubbles. Because of the low viscosity of the glass, these bubbles rise quickly and take other small gas inclusions with them, thus effecting an additional homogenization of the glass melt.

From the purification chamber the glass flows into the stirring crucible, where the glass temperatures are reduced to approximately 1400 °C. As a result, any small gas inclusion in the glass melt is reabsorbed. In addition, the glass melt is further homogenized by a stirrer. The glass melt flows through an electrically heated noble-metal pipe (feeder) to the hot-forming process, whereby the glass temperatures are lowered to approximately 1350 °C.

Discontinuous Melting

The discontinuously operated melting unit (see Fig. 4.13) consists of the following components: melting tank (melting volumes up to 28 m^3), insertion system, arch heating, electrical bottom heating, feeder, cooling system, waste gas disposal, and instrumentation and control with process computer.

In the discontinuous melting process, the individual manufacturing steps previously described do not take place at different locations, but are carried out one after the other. The batch is placed into the empty melting tank with a shovel and melted by arch heating as well as by electrical bot-

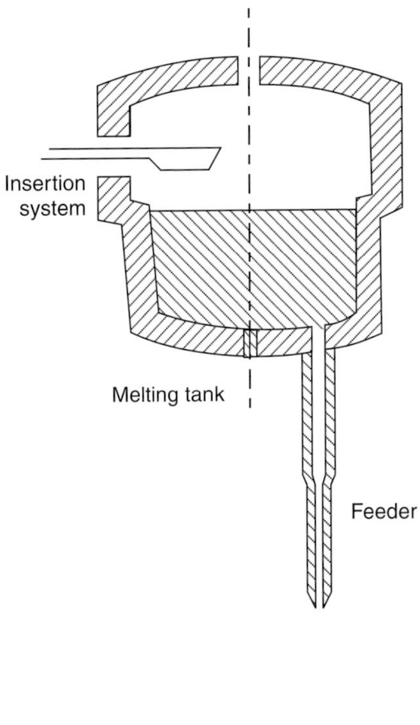

Fig. 4.13. Schematic representation of a discontinuously operated melting unit

tom heating after the bottom electrodes have been covered. For example, the insertion process lasts approximately ten days for a melting tank with a volume of 28 m^3. The so-called refining phase takes place afterwards over a period of about five days, for which the temperature of the melt is raised to approximately 1600 °C. As described above, refining bubbles are produced, which transport small gaseous inclusions with themselves and also homogenize the melt. During the refining phase as well as during the subsequent so-called cooling phase, the glass is further homogenized. During the cooling phase (duration approximately five days), the temperature of the glass melt is reduced to about 1400 °C. During the casting the glass flows through the electrically heated feeder into the casting mold. With large-volume cast pieces, throughputs of up to 300 kg/min can be reached.

4.2.2 Hot Forming

After leaving the feeder, the hot material is shaped into different forms. As previously described, throughputs of up to 1000 g/min can be achieved with continuously operating tanks, whereas with discontinuously operating tanks throughputs of up to 300 kg/min can be obtained.

Bars

The glass for the manufacturing of bars is melted continuously. Bars with cross-sections from 120 mm × 20 mm to 180 mm × 40 mm are manufactured with a bar-drawing device (Fig. 4.14). The most important components of this device are: shaping part with heating and cooling unit, drawing unit, annealing lehr, and bar cutting unit.

Bar manufacturing is a continuous horizontal strand-drawing process. The melted glass flows out of the feeder into the open, U-formed shaping part and is pulled out by means of a drawing device. The drawing speed and the throughput determine the thickness of the rectangular glass bar with a given width of the shaping part. The cutting of the bar to the required delivery length takes place either at the end of the drawing part or after the bar has

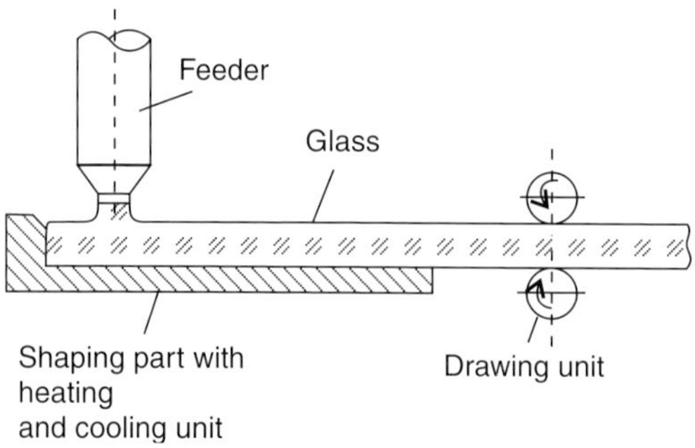

Fig. 4.14. Schematic representation of a bar drawing device

been cooled to almost room temperature in an annealing lehr. For cutting, the bar is scribed on one side and then broken by applying a force.

Blocks

The glass for the manufacturing of blocks is melted continuously. Blocks with dimensions from $180 \times 180 \times 180\,mm^3$ to $320 \times 320 \times 180\,mm^3$ are manufactured on a fully automatic casting machine (Fig. 4.15).

The device consists of the following main components: two casting/pressing stations, automatic continuous lowering of the mold during filling, automatic strand cutting fixture, and instrumentation and control.

The melted glass flows continuously out of the feeder. The glass-strand cutter cuts the glass strand and holds it until the casting/pressing station has moved under the feeder. The strand cutter then lays the glass strands onto the bottom of the casting mold. During the filling of the casting mold, the mold is lowered in such a way that the distance between the feeder and the glass surface remains constant. After completion of the filling process, the strand cutter cuts the glass strand and keeps it until the pressing station is driven out. The cutter opens afterwards. While the second casting station is driven in as previously described, the glass block cools down in the casting mold. When the glass has reached a temperature at which the block is no longer deformed plastically, it is fed into an annealing oven with a manipulator and cooled to room temperature.

Round Disks

In the production of round disks with a diameter from 500 mm to 4400 mm and a thickness from 200 mm to 1400 mm, blanks of the largest possible volume are manufactured including the stage of the ceramizing process. From these blanks the desired dimensions are manufactured by sawing, grinding,

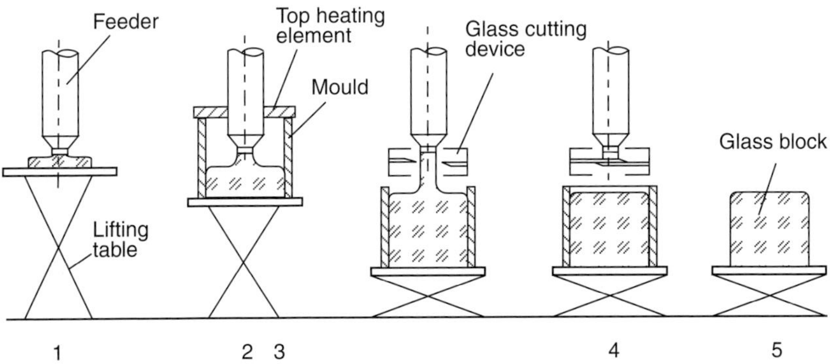

Fig. 4.15. Schematic representation of a fully automatic casting machine

boring, etc. For blanks with dimensions up to 1300 mm in diameter and 300 mm in thickness, the glass is melted continuously, whereas for larger dimensions it is melted discontinuously.

The most important manufacturing components are: casting mold, top-heating device, lifting platform, and stopping fixture for the glass strand.

The casting mold consists of a support structure, the mold bottom, and the mold wall. The mold bottom and the mold wall are covered with a parting compound on the glass side. The parting compound protects the mold bottom and wall from excessive heating, ensures the formation of only a thin crystal layer (see Sect. 4.3) and, furthermore, guarantees that the blank can be separated from the casting mold after the coarse annealing.

The casting mold is covered with the so-called top-heating device with which it is preheated to approximately 1000 °C before casting.

With smaller sizes (continuous melting), the glass is fed into the casting mold in the same way as is done for the manufacturing of blocks.

For large-volume sizes, the inner pipe of the feeder is heated through resistance heating, so that the glass flows out of the feeder. The glass flowing out of the feeder at the beginning is only used to purge the feeder; it cannot be used for the casting itself. After having reached an adequate quality, the melted glass is fed into the casting mold. This procedure is the same for small and large blanks.

It is of vital importance that the distance between the lower edge of the feeder and the glass surface is set at a defined value. The striation and the bubble quality are decisively affected by the maintenance of a defined spacing. For this purpose, the casting mold is continuously lowered with a lifting platform.

With a discontinuous melt, the glass strand is stopped by a suitable fixture after completion of the casting process.

4.2.3 Annealing and Ceramizing

For logistical reasons, the cooling to room temperature and the conversion to glass ceramic take place in two stages. After cooling, it is possible to remove glass faults from the workpiece. This definitely lowers the risk of breakage in the ceramizing, meaning that the yield is increased.

Annealing

After casting, the bar, block, or round disk is cooled to room temperature in the glassy condition. This process consists of two phases (Fig. 4.16):

- rapid cooling to T_g,
- fine annealing.

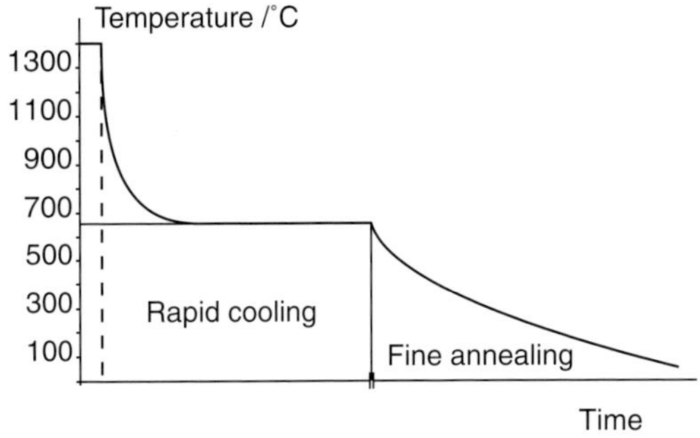

Fig. 4.16. Schematic representation of an annealing program for a round Zerodur® disk

First rapid cooling to T_g: during the cooling phase, Zerodur® tends to severe devitrification at temperatures between 1200 °C and 900 °C, especially at the contact surface with the parting compound. The resulting crystal layer has a lower coefficient of thermal expansion than the glassy material. During the cooling, the glass thereby shrinks on the crystal layer. Below the solidification temperature (second cooling phase), the glassy material is subjected to tensile stress. Cracks can be produced in the glass, which can lead to rupture when the glass is cooled to room temperature. The stresses formed depend upon the thickness of the crystal layer.

To reduce the thickness of the crystal layer, the temperature range between 1200 °C and 900 °C must, therefore, be crossed as quickly as possible. This is achieved by allowing the glass to radiate freely upwards until a core temperature of about 900 °C is reached, whereby an inhomogeneous temperature distribution is produced in the glass. The surface is up to 250 °C colder than the core. Attention must be paid to the surface temperature not being significantly less than 600 °C. Otherwise, cooling cracks can be produced at the surface. At a surface temperature of 600 °C, the blank is, therefore, to be placed into an annealing oven.

As mentioned above, the temperature distribution in the blank is homogenized at T_g, meaning that the temperature of the surface is raised to 675 °C. During this phase it is essential that those zones of the blank that have reached temperatures below 740 °C are never reheated to temperatures above 800 °C. Nuclei are produced below 740 °C and crystals can grow on them when the blank is reheated (particularly above 800 °C). A partially crystallized zone results, which is different from the other material with respect to its thermal expansion coefficient. Because of this difference in thermal expansion, stresses that can lead to rupture of the glass may form during cooling to room temperature.

Second fine annealing: according to the rate cooling process, a certain temperature distribution is established in the blank until the solidification temperature is reached. It determines the permanent stress (σ) after annealing at room temperature;

$$\sigma = \frac{E\alpha}{1-\mu}\frac{v}{\lambda}F \ , \qquad (4.1)$$

where E is the elastic (Young's) modulus, α is the thermal expansion coefficient, μ is Poisson's ratio, v is the cooling rate, λ is the thermal conductivity, and F is a geometrical factor (see also another volume of this series: *The Properties of Optical Glass*, Chap. 5 [4.36]).

The cooling rate in the solidification range is to be set in such a way that the permanent stresses are low enough to allow the machining of the blank.

At temperatures beneath the solidification range, it must be ensured that no temporary critical tensile stresses, which could lead to rupture of the blank, can occur at the surface because of a change in the temperature distribution. Temporary tensile stresses always occur at the surface if the quotient v/λ increases below the solidification temperature.

In cast pieces with damaged surfaces in the area of the crystal layer, the temperature at which the rupture of the cast pieces can occur is low. For this reason, the cooling rate must be reduced with decreasing temperature in such a way that the reduction of the thermal conductivity is compensated or, better, overcompensated. Through overcompensation, a part of the permanent compressive stress at the surface of the blank is already established during the coarse annealing. Thus, the probability of rupture is reduced.

Ceramizing

After rough machining (and the working out of faults in the glass, if required), the blanks are converted from the glassy condition into the semi-crystalline condition through a thermal treatment. Due to that treatment the material also receives the property of zero thermal expansion.

In general, the ceramizing is carried out in normal annealing lehrs. The location of the blank in the annealing oven and its support are selected so that a good temperature distribution is guaranteed in the glass. As shown in Fig. 4.17, the ceramizing process consists of four phases:

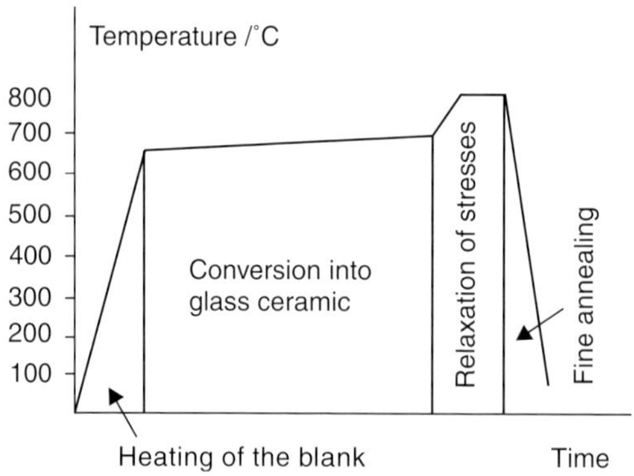

Fig. 4.17. Schematic representation of a ceramizing program for a round Zerodur® disk

- heating of the blank,
- conversion into glass ceramic,
- relaxation of the stresses produced during the conversion and establishment of the defined coefficient of thermal expansion,
- fine annealing.

The blank is heated slowly to a temperature in the vicinity of T_g without the occurrence of critical stresses.

The conversion into glass ceramic is an exothermic process. Energy amounting to 250 kJ/kg is released. Furthermore, the material shrinks by approximately 3 vol% during the conversion. To avoid critical stresses, the conversion process must take place as uniformly as possible throughout the entire volume of the blank, for which the temperature must be established very homogeneously in the blank (e.g., $\Delta T \leq 2\,°C$ with large-volume blanks). The required temperature homogeneity can be obtained only when the heat-up rate is so low that the heat produced in the exothermic process can be transported to the outside of the blank. As a result of this, the conversion can last up to nine months for large blanks.

The diagram (Fig. 4.18) shows the viscosity of the material as a function of temperature during the conversion process for a particular ceramizing program. During the conversion phase, the viscosity rises in such a way that the stresses produced cannot relax completely. Only after a further temperature increase is the viscosity diminished to a level which allows the extensive reduction of the stresses.

With an increasing dwell time at high temperatures, a slight change of the crystal modification raises the thermal expansion coefficient. Because the material composition is prescribed in such a way that the thermal expansion coefficient is slightly negative after the conversion phase, it is possible to

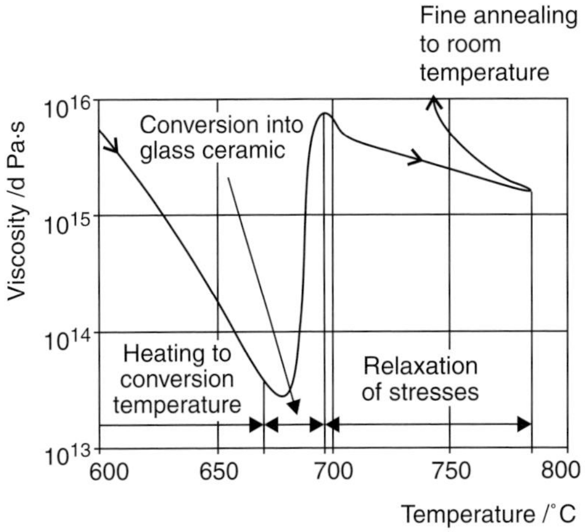

Fig. 4.18. Viscosity as a function of temperature during a conversion process

establish a definite "zero expansion" condition by selecting the holding time at a certain temperature.

The establishment of a defined annealing bulk stress takes place as described in the above part on annealing.

4.2.4 Machining

The processes, machines, and tools used in the machining of Zerodur® are identical to those used for optical glass. Zerodur® can be ground, lapped, and polished; the drilling of through-holes and blind holes, as well as the preparation of filigree structures, for example, by grinding, ultrasonic drilling, or water jet cutting, are practicable without any problems.

Grinding is performed by diamond tools, where countless diamond grains of a determined size are incorporated in a metal or plastic matrix. Electroplated tools (mainly special shaped tools) are used likewise. The type of machining aimed at determines the grain size to be chosen: roughing processes call for a coarse grain (D 251, e.g.), while ready-for-lapping or even polishable surfaces are obtained by using grain sizes around D 64 and D 7, respectively. Investigations of sample surfaces ground with the aid of D 64–D 251 have produced peak-to-valley heights (R_t) between 30 and 50% of the grain sizes used and microcrack depths between 60 and 120%.

The type of machining determines the tool shapes. Circumferential grinding wheels are customary for circular grinding, but likewise for spherical surfaces. Cup wheels are used for the working of flat and spherical surfaces, core drills for through-holes or for the production of rods, as well as various types of special shaped tools.

Lapping is the machining – with loose abrasive – by steel or greyiron tools formed to fit the final shape. Silicon carbide of appropriate grain sizes is frequently used as abrasive powder. Properly speaking, sawing is also a grinding process using loose abrasive (wire saw), or bonded diamond grain (saw blade). Looked at from outside, polishing seems to be very similar to lapping. Both need a tool, normally with precisely the contour of the workpiece, but as a negative. This tool is covered with a felt or plastic layer (frequently polyurethane); as a rule, cerium oxide slurries are used as polishing agents. Polishing produces excellent surface finishes on Zerodur®. Surface residual roughness is less than 4 Å in the case of the ROSAT X-ray telescope mirrors polished by Carl Zeiss, Oberkochen.

Zerodur® has marked advantages in the polishing of precise workpieces because the material does not deform upon process-associated heat-up; consequently, delays between the individual measuring and machining cycles are only slight.

4.2.5 Quality Assurance

All testing and control steps are integrated into the manufacturing process. The elements of the quality assurance system are designed according to ISO 9001. Thus, it is ensured that the manufactured products comply with specifications. This is demonstrated by a process such as the one specially used for 8-m class mirror substrates. Correspondingly, the process is less complicated for large-scale Zerodur® products than for blocks and bars with smaller dimensions.

Raw Materials, Glass Batch

The analyses of raw materials and the performance of test melts are aimed at:

- determining the content of the basic components, and
- recognizing discoloring impurities or altered melting behavior in due time.

Melting

During the melting process, regular tests are performed on cast samples. Glass samples are examined regarding:

- control of internal quality (bubbles, inclusions, striae),
- analysis of the glass components.

Furthermore, specimens are subjected to temper programs for fast ceramization, in order to gain insight into the properties of the subsequent glass ceramic during the process:

- control of the physical properties such as density, crystal size, crystal content, etc.,
- control of the expansion behavior.

The frequency of the test is determined by the course of the process. Possible corrective measures modify the glass batch and the process control during melting. For large pieces Zerodur® is melted discontinuously. If the above-mentioned tests reveal that the necessary glass quality has been attained, casting approval is given. For smaller dimensions (bars, blocks), on the other hand, Zerodur® is melted continuously.

Within the scope of producing 8-m mirrors, special tests have been conducted during the hot forming and annealing process stages.

Hot Forming

This involves:

- contour check of the casting mold with a theodolite system for producing near to net shape and, thus, "save" glass,
- inclusion detection in hot state for an evaluation of the internal quality.

Annealing

The observations of the blank in the annealing furnace (growth of crystal skin, shell formation, depressions) from many individual annealing processes leads to optimized processes and a high level of process reliability.

Raw Glass Machining

After the removal of the crystal skin and the possibly conchoidal bottom, informative testing of the internal quality is possible on the upper side which has been fire-polished. The following tests are conducted upon this raw glass treatment (goal: final determination of inclusion quality, preliminary information about stress birefringence at inclusions and striae, and about bulk stress).

- Final detection of bubbles and inclusions with a diameter of > 0.1 mm. Depending on the specifications, the diameter and position of inclusions are determined with a microscope.
- Measurement of stress at all inclusions (Senarmont).
- Testing for striation (stress-optical); if, due to the usage, optical quality in transmission is required, the shadow method is applied to detect striae.
- Detection of all surface damages.
- Determination of the usable volume or preliminary position of the blank in the raw casting. When using reflective optics (as opposed to the previously mentioned transmission application) the critical volume over the position of the application surface (optically effective surface) is determined for the evaluation of the glass flaws in the specification. The internal quality is more strictly specified in this critical volume.

Ceramization

Specimens are taken from the blank within the scope of raw glass machining (rods with a length of 100 mm and a thickness of approximately 5 mm) in accordance with the specifications. These specimens are laid out near the blank in the annealing furnace as specified in the testing plan. Using these specimens, the spatial distribution of the expansion coefficient is determined and, therefore, the spatial consistency of the ceramization process proved. After ceramization, a global stress test (homogeneous stress distribution – no light patches) and a local stress test (low bulk stress – low compressive stress on all surfaces) are conducted.

Final statements on the stresses at inclusions and striae, on the bulk stress and the CTE can now be made.

Based on the overall evaluation of the raw casting, the position of the blank in the raw casting can be finally determined. A detailed machining plan is drawn up.

Final Machining

The final product compliance with specifications is ensured by means of intermediate controls (dimensions, contour measurements, surface quality), which are followed by releases for further machining on the part of quality assurance.

Documentation

Following the final control, all of the features called for in the specification are compared with the measured values. Tracing from the glass batch to the finished product is possible using the melt number and all annealing numbers.

4.3 Production of Zerodur® in Special Shapes

Hartmut Höness, Alfred Jacobsen, Konrad Knapp, Thomas Marx, Hans Morian, Rudolf Müller, Norbert Reisert, Armin Thomas

For certain uses of Zerodur®, blanks with special geometries are to be manufactured, such as, for example, blanks with large volumes and/or extremely thin walls. For this purpose, special and/or additional manufacturing processes and equipment as compared to conventional production (see Sect. 4.2) must be used. The development tasks and the solutions for the manufacturing are described in the following.

4.3.1 Thin Menisci

Thin menisci made of Zerodur® are used as mirror substrates in astronomical telescopes (see Sect. 4.6.1). For massive conventional mirror substrates, a thickness:diameter ratio of 1:6 was a usual rule. With this ratio, the mirror blank is adequately stiff. This means that its function is not restricted by deviations in the contour caused by weight, wind pressure, temperature inhomogeneities, etc.

This rule is no longer applicable when the following requirements must be met:

- With the help of an active support system, the geometry of the mirror blank is changed in such a way that its contour is as close to the ideal optical contour as possible.
- The diameter of the mirror blank is so large that the application of the above rule would result in volumes that could hardly be managed technically and in manufacturing costs that could no longer be financed.

In both cases, the thickness:diameter ratio is approximately 1:20 to 1:50. Such mirror substrates are called thin menisci.

Production Techniques

There are three essentially different techniques for the manufacturing of thin menisci from Zerodur®:

- the conventional technique,
- the slumping technique,
- the spin casting technique.

In the *conventional technique* (see Sect. 4.2.2), thin plates are cut with a wire saw from thick, already ceramized cast pieces and are ground with diamond tools on all sides to achieve the desired contours, thicknesses, and diameters.

Because the curvature can be produced only by grinding, the material usage is very high so that this process can only be taken into consideration for menisci with low camber in comparison with the thickness and for individual manufacturing.

The *slumping technique* also uses plates cut from thick glassy cast pieces, but they have a thickness which is already very close to the final thickness. These plates are sunk into a concave mold from above with pressure applied on one side through plastic deformation above the transformation temperature. The conversion to glass ceramic takes place after the slumping process. The final geometry is established through subsequent machining.

Because of the additional investments (mold, oven, loading fixture), this process is suitable only for large series. The thickness is presently restricted to a maximum of approximately 100 mm.

In the *spin casting technique*, the melted glass is poured into a mold with a concave bottom facing upwards. After pouring, the concave upper side of the meniscus is generated through rotation of the casting mold. After reaching the required surface contour, the glass is cooled while the rotation is maintained, so that the contour is retained. The blank is then – as with conventional casting – cooled to room temperature. After rough machining, the conversion to glass ceramic takes place.

With the slumping technique and the spin casting technique, the material consumed can be reduced drastically in comparison with the conventional technique, especially for workpieces with a large camber. Figure 4.19 shows the most economical manufacturing techniques for different diameters, thicknesses, and apertures.

Spin Casting of Thin Menisci with Large Volumes: 8.2-m Zerodur® Mirror Blanks for the ESO/VLT

Between 1993 and 1996, Schott manufactured and delivered four 8.2-m mirror blanks, with a thickness of 177 mm and a radius of curvature of 28,975 m, for the Very Large Telescope (VLT) of the European Southern Observatory (ESO).

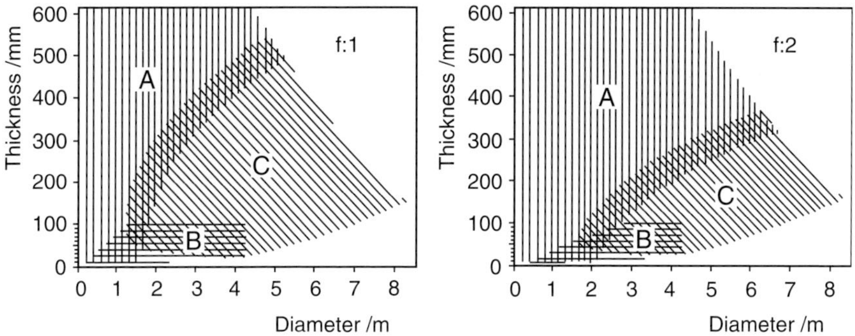

Fig. 4.19. The most economical manufacturing techniques for thin menisci for two relative apertures (f:1, f:2): (A) conventional technique, (B) slumping technique, (C) spin casting technique

Setup of the Manufacturing Equipment. To fulfill the ESO order, all buildings and equipment for the manufacturing were newly built between September 1988 and March 1991. The industrial harbor site (approximately 400 m away from the main works in Mainz) was selected because of its outstanding transportation connections: the Rhine harbor, a four-lane motorway, and a railway.

The buildings (built-up area approximately 50 000 m^3) are arranged as follows (Figs. 4.20 and 4.21).

- R 1: cooling and ceramizing facilities for mirror substrates up to 4.4 m diameter.
- R 2: medium supplies, cooling systems, mechanical workshop.
- R 6: office and personnel rooms, electrical supply.
- R 7: spin casting facilities (with top heating device and cooling cover for blanks), three cooling and ceramizing ovens, two mold construction stations, one storage area with three storage containers, grinding machine with tempering cabin, quality test stand, handling facilities (suction lifter and turning device), 70 t crane.
- R 8: Central control room for process control, 70 t melting tank, insertion system, feeder, arch heating, electrical bottom heating, tank cooling, waste gas disposal.

Survey of the Manufacturing Cycle. The spin casting technique is used for the manufacturing of the mirror substrates. The most important manufacturing steps are:

- melting of the glass in a 70 t melting tank,
- spin casting (pouring of the melted glass into the casting mold and spinning),
- coarse annealing to room temperature in the glassy condition,
- handling (transport with suction lifter and turning),

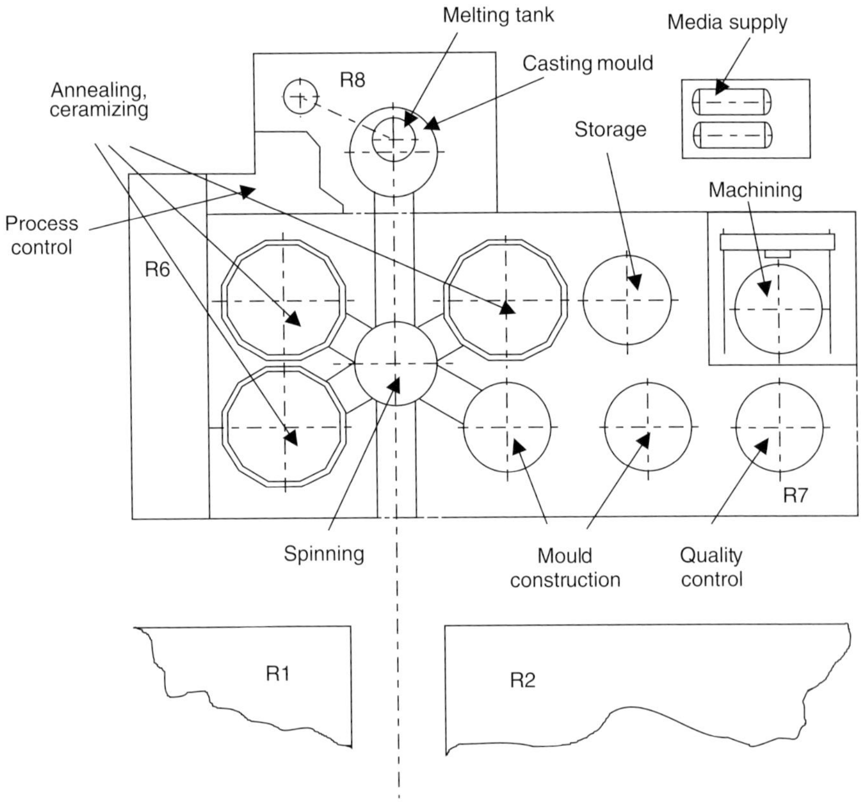

Fig. 4.20. Top view of the manufacturing facility for the 8-m mirror substrates

- rough machining,
- handling (transport with suction lifter),
- ceramizing (conversion to glass ceramic),
- handling (transport with suction lifter),
- manufacturing of the center hole,
- handling (transport with suction lifter),
- fine annealing (establishment of a defined stress distribution through thermal post-treatment),
- handling (transport with suction lifter and turning),
- final machining,
- final quality control,
- handling (transport with suction lifter to the shipping container)

The total manufacturing time is two years (quality controls in-between the individual manufacturing steps are not mentioned).

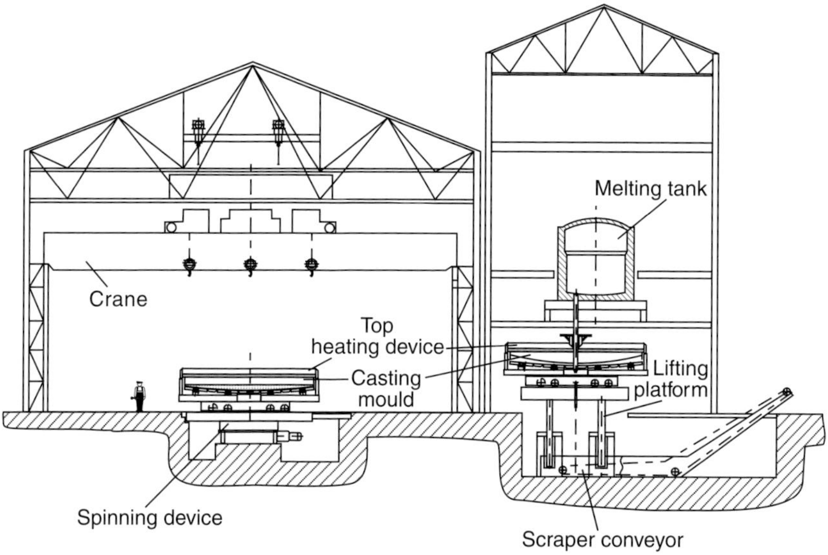

Fig. 4.21. Side view of the manufacturing facility for the 8-m mirror substrates

Discontinuous Melting

The quality of the glass to be fed into the mold had to be improved for the manufacturing of the 8-m mirror substrates. In the spin casting technique, there is no longer the possibility to avoid glass faults such as bubbles, solid inclusions, crystals, and striae in the final machining, as there is with the conventional technique.

The glass is melted discontinuously, as described in Sect. 4.2.1, in a melting tank with a volume of 28 m^3 (corresponds to approximately 70 t of glass) and complies with the specification about 20 days after batch feeding. Using such large-volume melting tanks, the high quality requirements can be met or even surpassed. For instance, the permissible number of inclusions breaking into the polished surface of the first VLT 8.2-m mirror blank was $< 0.5\,m^{-2}$; this value has been exactly achieved. The specification for stress birefringence caused by striae was < 25 nm, the actually achieved value is 0.

Spin Casting

As explained in Sect. 4.2.3, the glassy Zerodur® tends to devitrify at the contact surface with the parting compound of the casting mold. During the cooldown process, this crystal layer leads to tensile stresses in the glass because of the different rates of thermal expansion of the glass and the crystal layer. The so-called "meniscus effect" (see the paragraph on "Annealing" on this) causes the stresses to increase locally in thin menisci. In blanks with surface damage these stresses can lead to rupture. For this reason, the pouring and spinning process are to be carried out in such a way that sufficiently thin crystal layers are produced, which do not lead to critical stresses.

The spin casting process can be divided into the pouring part and the spinning part.

Pouring Process

The most important manufacturing components for the pouring process are: the casting mold, top heating device with video systems, transport unit, lifting platform, device for separating the glass strand under the casting mold (called the lower shears in the following), device for separating the glass strand over the casting mold (called the upper shears in the following), plug for stopping the glass strand, scraper conveyor to dispose of the purging glass, and various instrumentation systems. The devices are presented schematically in Fig. 4.22.

Description of the Function of the Casting Mold. As shown in the schematic representation (Fig. 4.23), the casting mold consists of a rigid base structure, 18 movable star-shaped supports, ribs between the star-shaped supports, segmented bottom plates above the star-shaped supports and the ribs, and a segmented mold wall. The bottom plates and the wall plates are covered with a parting compound.

The mold material is a special steel which can withstand several temperature cycles without subjecting the individual elements to unacceptably large deformations or changing the properties of the material unacceptably.

The 18 star-shaped supports can be raised using pneumatically operated metal bellows. During the pouring and spinning process, however, the metal bellows are not under pressure. The star-shaped supports rest on the base structure only at their center and by threaded rods fixed at their edge. During the plastic condition of the glass, the contour of the blank is prescribed by the rigid casting mold. The functions of the casting mold for the visco-elastic and the elastic conditions of the glass are described in the paragraph on "Annealing".

Manufacturing Cycle. Before casting, the casting mold is preheated with the top heating device. The casting mold is then transported to the pouring station with the transport vehicle and lifted there with the lifting device, whereby the feeder is guided through a hole in the top heating device and a hole in the bottom of the casting mold.

The inner pipe of the feeder is heated by a resistance heater so that the glass runs out of the feeder. The glass flowing through the feeder at the beginning is used to purge the feeder. It falls into a scraper conveyor, in which it is granulated and then disposed of. When an adequate glass quality is reached, the lower shears are driven in. The plate of the lower shears is scoured, heated, and cleaned by the melted glass, after which the plate of the

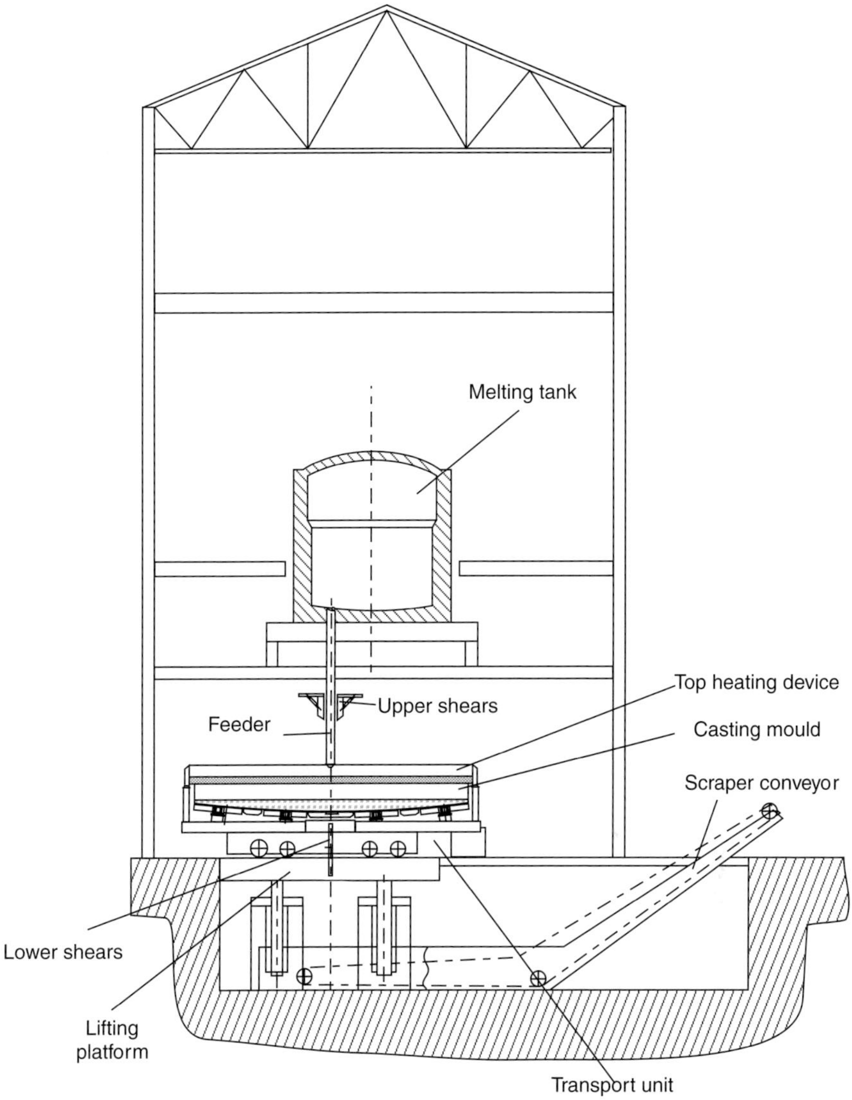

Fig. 4.22. Schematic representation of the devices for the casting of 8-m blanks

lower shears is pressed into the hole of the mold bottom. The glass strand is cut off at the edge of the shears plate.

After this, the casting mold is filled with the melted glass. The distance between the lower edge of the feeder and the glass surface is observed through video systems and can thereby be set to a defined value. The casting takes

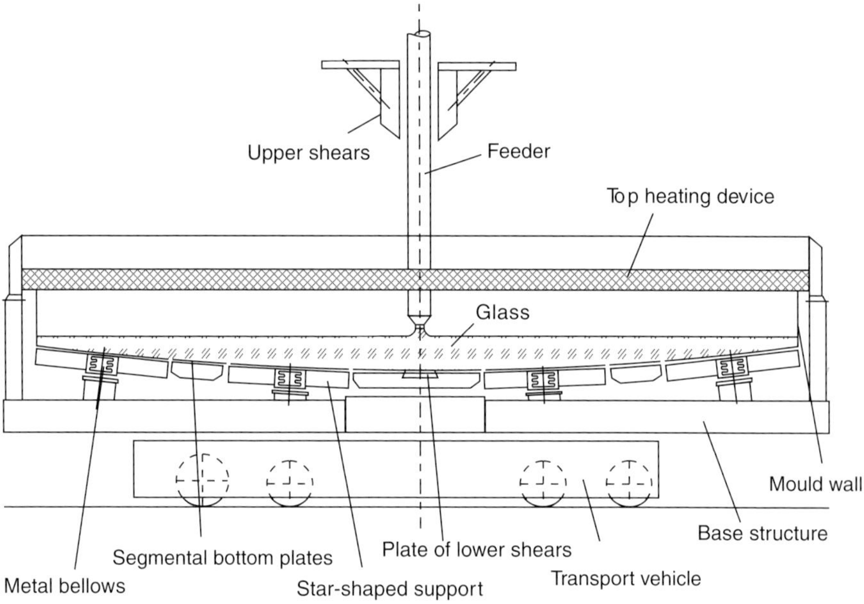

Fig. 4.23. Schematic representation of the casting mold

place with an average flow of approximately 200 kg/min, which means that the pouring time amounts to about four hours.

At the end of the pouring process, the glass strand must be stopped, for which the casting mold is lowered. Synchronized with the lowering of the casting mold, the so-called upper shears are driven into the casting mold with the shear halves opened. The two shear halves are closed and the glass strand is stopped between the feeder and the upper shears. This prevents the glass strand from penetrating into the glass with an excessively high speed and bringing bubbles and striae with it. After the glass strand has been stopped, the filled casting mold is transported to the spinning position with a vehicle and fixed there.

Minimization of the Crystal Layer Thickness. As mentioned above, the thickness of the crystal layer, which is formed at the contact surface with the parting compound during the pouring and spinning process, is to be minimized. The thickness of the crystal layer depends on the parting compound used, the contact temperature between the glass and the parting compound when pouring, the pouring duration, and the cooling duration.

Figure 4.24 shows the relationship between the thickness of the crystal layer and the contact temperature (pouring time 4 h) for different parting compound compositions (A, B, C).

It can be seen clearly that optimization of the parting compound composition and optimization of the temperature control during the pouring can

(a) the deformation of the blank under its own weight on the support system in the kiln,
(b) the changing of the temperature differences in the blank during the cooldown (annealing bulk stresses),
(c) the different shrinkage behavior of the glass and the crystal layer on the bottom of the blank during the cooling.

Stresses Caused by the Support System. After dropping below the transformation temperature, the glass is in a visco-elastic or a brittle-elastic condition. It must then be observed that the 8-m blank has a limited stiffness because of its low thickness. This means that, because of the deformation of the 8-m blank under its own weight, tensile stresses of $\geq 5\,\mathrm{N/mm^2}$ occur at the surface with a maximum deformation of 0.8 mm with a static three-point support. In particular, blanks with damaged surfaces can break due to tensile stresses of this magnitude.

As a result, the mirror blanks must be supported in such a way that no critical tensile stresses occur (objective: tensile stress $\leq 1\,\mathrm{N/mm^2}$) during all production steps in which the glass is in a visco-elastic or elastic condition (annealing, ceramizing, transport, turning, machining, and quality control).

The blank must be supported over a large area and the support must adapt to the contour of the blank, whereby it must be taken into account that the contour of the blank as well as the support system change during the manufacturing sequence.

In the visco-elastic and the brittle-elastic conditions, the blank no longer adapts to the contour of the mold bottom as it does in the viscous condition. Although the contours of the reverse side of the blank and the casting mold are identical at the beginning of the cooling process, they deviate more and more from each other with increasing cooling time because of the different shrinkage behavior of the glass and the steel structure. To avoid critical stresses, the support system must, therefore, adapt to the contour of the blank.

After reaching the transformation temperature, the 18 metal bellows of the star-shaped supports are pressurised (see the paragraph on "Pouring Process" for details of the mold structure). The pressure is adjusted in such a way that only a part of the weight of the blank can be borne by these bellows. At the beginning of the cooldown process, in the visco-elastic range of the glass, therefore, the blank adapts to the contour of the casting mold by plastic flow. The metal bellows do not extend thereby. With progressive cooldown, the viscosity of the glass rises so that the blank can no longer adapt to the changing contour of the steel mold. By extending some of the metal bellows, the mold bottom adapts to the contour of the blank and these bellows support part of the weight of the blank. The remaining weight of the blank prevents the other metal bellows from extending. The threaded rods of the corresponding star-shaped supports transfer the remaining weight to the base structure. In this way, the position of the blank in the oven is defined.

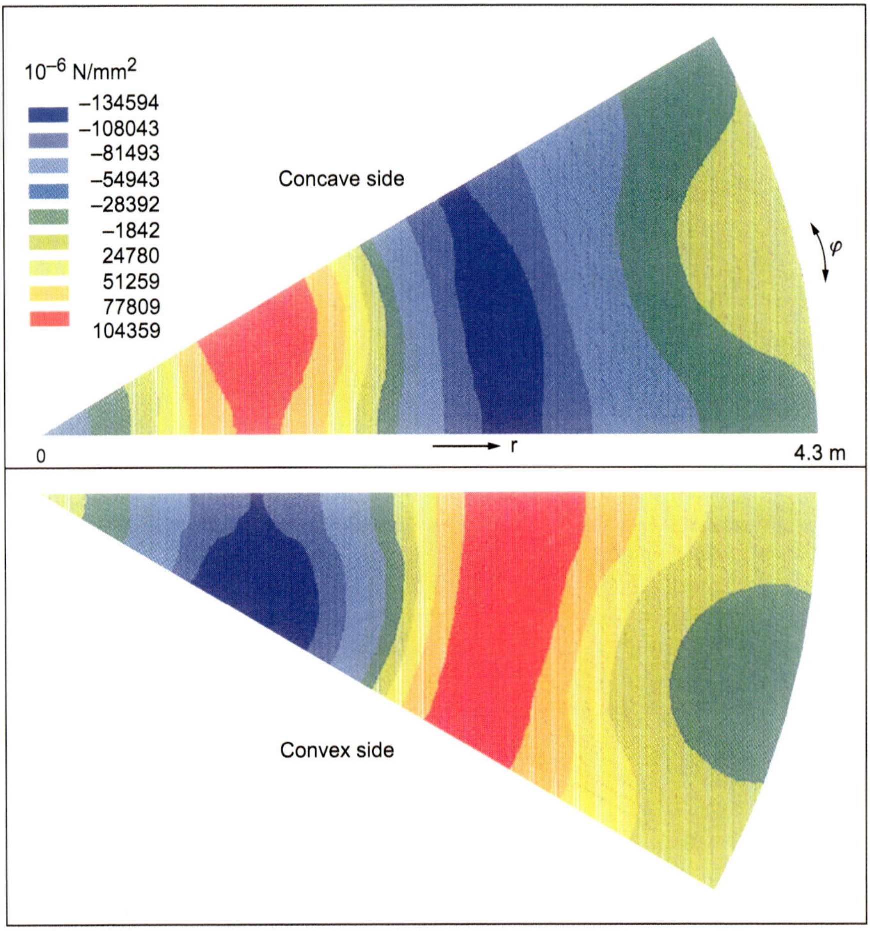

Fig. 4.30. Stresses in the r direction after the blank is cooled to room temperature

Figures 4.30–4.32 show the stresses of a blank, which were calculated with computer simulation (finite-element modeling). During the cooldown, the contour of the blank changes in comparison with the contour of the casting mold so that the 12 outer star-shaped supports have extended, while the six inner supports still rest on the base structure with their center and their threaded rods. The maximum tensile stress at the surface amounts to $0.13\,\mathrm{N/mm^2}$.

Annealing Bulk Stresses. As described in Subsect. 4.2.3, the annealing bulk stresses are established according to the rate cooling process. To reduce the probability of rupture, efforts are already made to produce compressive stresses on the surface of the blank during the annealing process.

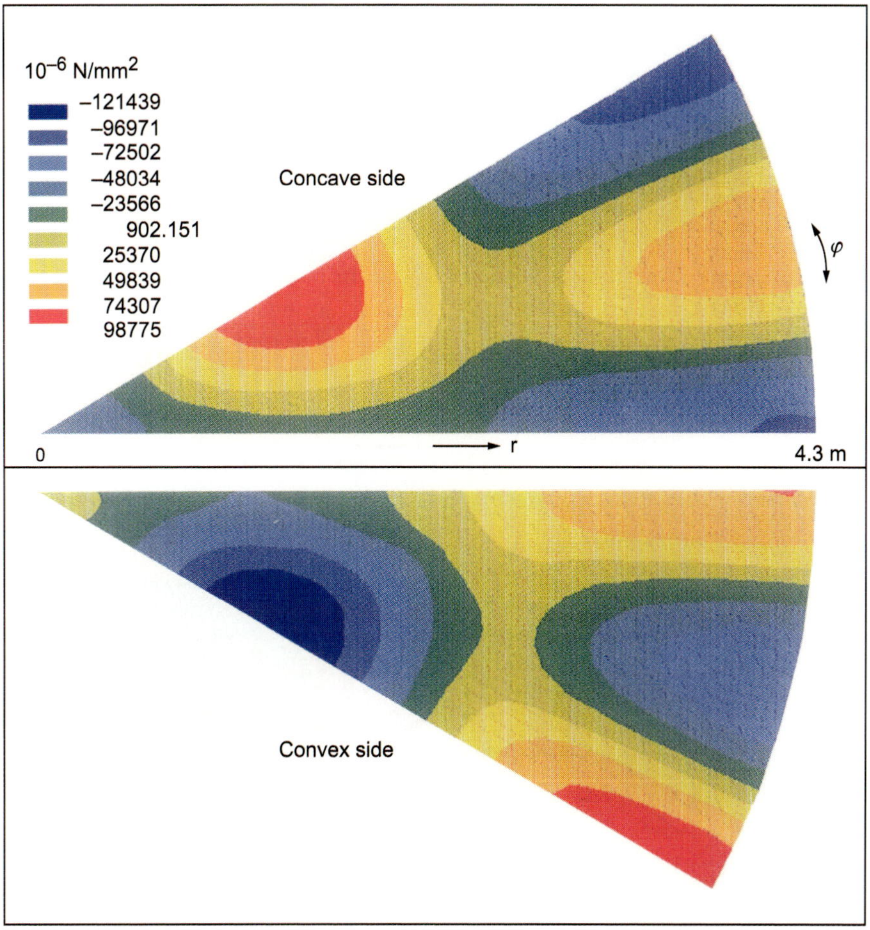

Fig. 4.31. Stresses in the φ direction after the blank is cooled to room temperature

The parting compound on the bottom and the edge of the blank leads to a non-uniform temperature distribution in the blank with cooling. Figure 4.33 shows the temperature field in the solidification range of the blank for a certain cooling program and a certain parting compound composition (determined with a finite-element calculation).

As shown in Table 4.1, the asymmetric temperature distributions in the thickness direction in a flat disk and a meniscus lead to very different stresses after cooling under otherwise equal boundary conditions. In particular, the compressive stress in the φ direction on the bottom is very strongly reduced in the meniscus in comparison with the flat disk. This means that the possibility to compensate for tensile stresses at the surface, which are caused by the crystal layer (see the following paragraph) or the support system, by

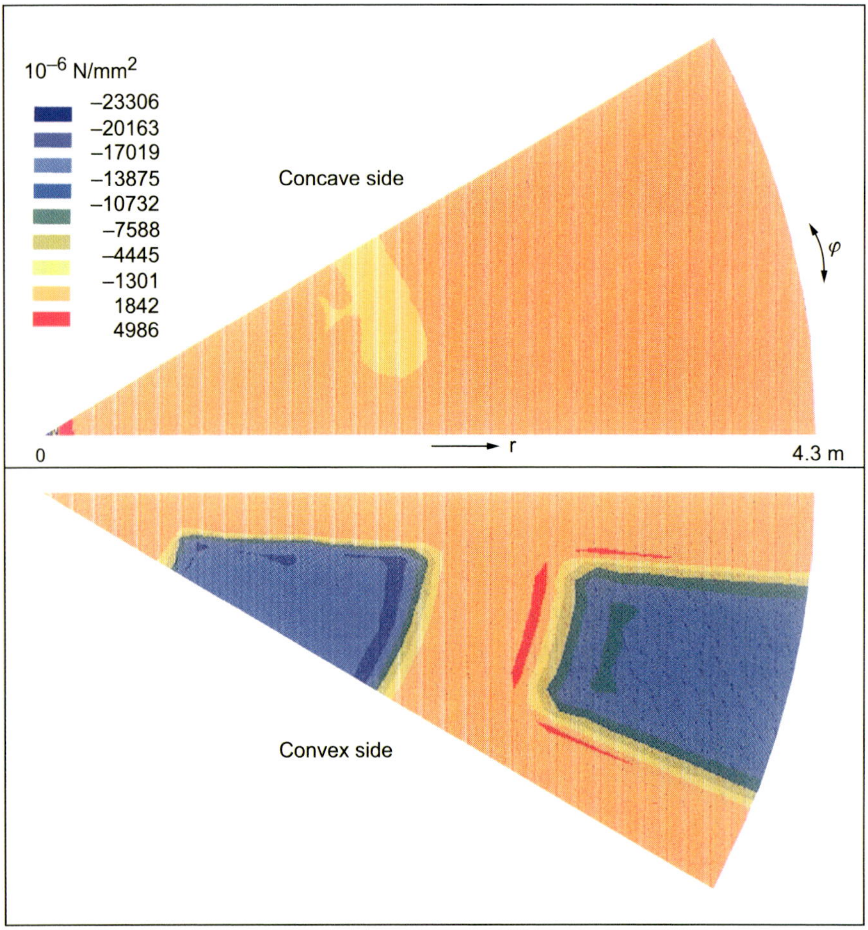

Fig. 4.32. Stresses in the z direction after the blank is cooled to room temperature

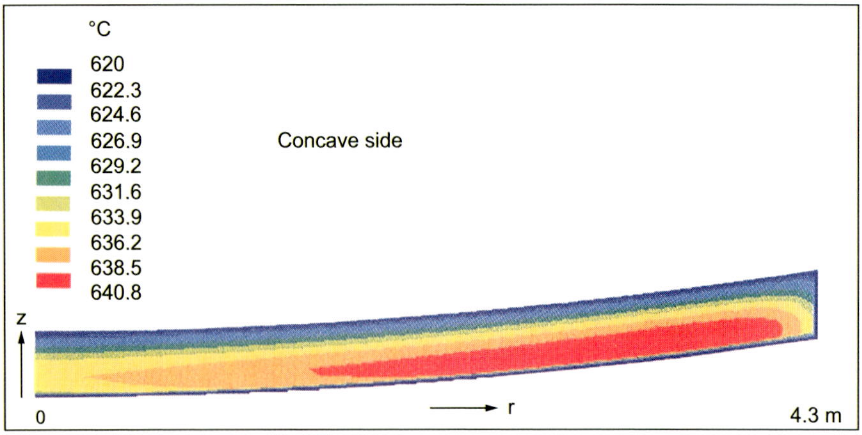

Fig. 4.33. Temperature field in the solidification area of the blank

Table 4.1. Stress due to asymmetric temperature distributions in a flat disk and a meniscus

Shape	Side	Stress N/mm^2			
		in φ direction		in r direction	
		center	edge	center	edge
Flat disk	top side	−2.5	−2.5	−2.5	−2.5
	bottom side	−2	−2.3	−2	−2
Meniscus	top side	−4.8		−4.8	
	bottom side	−1.2	−0.3	−1.2	−2.2

annealing bulk stresses is significantly less with a meniscus-shaped disk than with a flat disk (called the "meniscus effect" in the following).

Stresses Caused by the Crystal Layer on the Bottom and at the Edge of the Blank. When cooling, the crystal layer on the bottom and at the edge of the blank shrinks less than the glass. This means that the crystal layer is subject to compressive stress, while the glass is under tensile stress. Figures 4.34 and 4.35 show the results of a finite-element calculation for a certain thickness of the crystal layer and a certain cooling program (parting compound B see Fig. 4.24). The principal stresses in the r and φ directions in the glass at the transition to the crystal layer are shown in the center, at the half radius, and at the edge as a function of the cooling time. The results also include the influence of the annealing bulk stresses. This calculation is based on a visco-elastic model taking account of the heat radiation.

Tensile stresses of up to 2 N/mm^2 are reached in the transition zone to the crystal layer. With a tensile stress of this order of magnitude (critical stress), the crystal layer breaks off from the glass, resulting in cracks in the

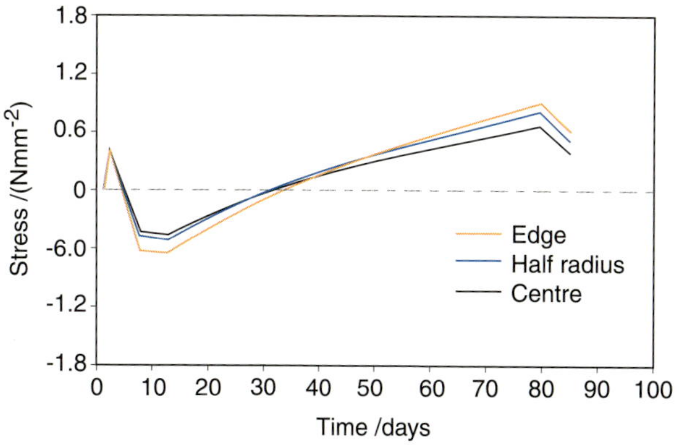

Fig. 4.34. Stresses in the r direction in the center, at half the radius, and at the edge as a function of the cooling time (bottom side of the blank)

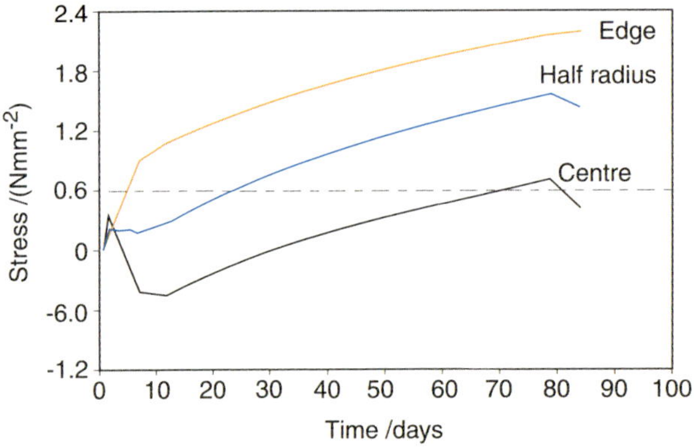

Fig. 4.35. Stresses in the φ direction in the center, at half the radius, and at the edge as a function of the cooling time (bottom side of the blank)

glass. The tensile stress is increased in the transition zone between the crystal layer still bonded and the crystal layer which has broken away (see Fig. 4.36). A separation of the crystal layer, therefore, takes place preferentially in this transition zone. This produces spherulitic or annular zones with a separated crystal layer. The separation of the crystal layer produces deep cracks and damage in the glass, which drastically increase the probability of rupture.

The parting compound C, see Fig. 4.24, avoids the separation of the crystal layer through crystal thicknesses ≤ 0.4 mm. With an optimum annealing procedure the critical stress values are not reached before room temperature.

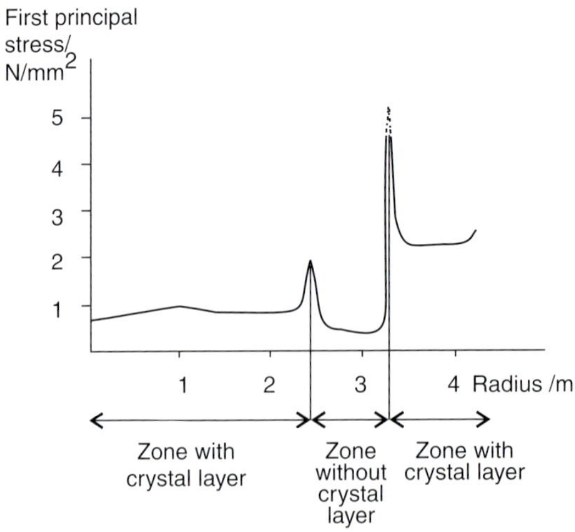

Fig. 4.36. First principal stress as a function of radius (bottom side of the blank, annealing temperature 350 °C)

Ceramizing

The temperature control for the conversion of glass into glass ceramic takes place as described in Sect. 4.2.3. There are special features, however, with the support of the blank during the ceramizing.

In the conversion, the blank changes its contour because of a volume shrinkage of approximately 3%. As previously described for the annealing process, the support system (modified casting mold) adapts to the contour of the blank through the extension of the metal bellows.

Local stress peaks, caused by non-uniform contact on the individual elements of the support system (e.g., deformation of the steel structure caused by plastic flow during the long holding time at high temperature) can lead to stress inhomogeneities after the cooling of the blank to room temperature. To avoid these stress peaks, the blank lies on a specially formed layer made of sand and mica. To equalize the temperature distribution in the thickness direction, the blank is also covered with an appropriate layer. The blank slides on the sand–mica layer when it shrinks during ceramizing. Because of its special geometry, this layer also has the characteristic that it equalizes pressure differences caused by the unequal support of the blank and changes in its geometry. It adapts locally to the contour of the blank or the support system. Figure 4.37 shows the contact pressure as a function of the filling height of the sand–mica layer. With uniform support of the blank the filling height of the sand–mica layer is 21 mm, and the contact pressure amounts to 0.006 N/mm^2. With a deformation of 0.5 mm, the contact pressure is doubled. With these low contact pressures, no stress inhomogeneities are to be expected after the ceramized blank is cooled to room temperature.

Handling and Support

As mentioned above the deformation of large thin menisci under their own weight is the main difference in the manufacturing steps following the casting as compared to the process for massive blanks. Tensile stress values of

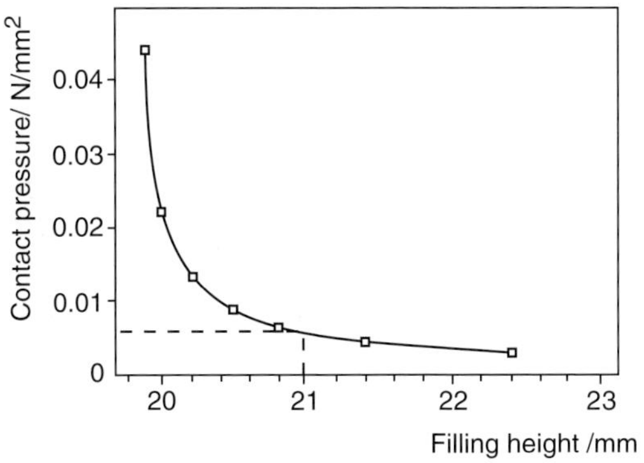

Fig. 4.37. Contact pressure as a function of the filling height of the sand–mica layer

about 5 N/mm^2 resulting from a static three-point support are beyond the permissible range since several of the handling steps will have to be carried out with the very rough casting surfaces on the edge and the bottom of the casting. Therefore, for all handling steps, like lifting, transport, setting down, turning, and storage, as well as for machining operations and quality control, devices had to be created which can always support the mirror blank so that the tensile stresses at its surface are not higher than 1 N/mm^2.

In order to fulfill this requirement the following concept has been applied to all pieces of equipment: the menisci have to be supported at multiple locations over a large surface area so that, at any time, there is a complete match between the curvature of the support system and the curvature of the blank. Since the blank as well as the support structure will deform during the different steps, the support structure has to be connected to the blank in a way which provides for independence of the above deformations from the transfer of forces between the blank and the structure. In other words, the numerous support elements have to exercise equal forces, which does not allow rigid connections between the blank and the support structure. This is achieved by using mechanical (whiffle-tree concept) or hydraulic elements between the support structure and the blank.

The above concept has been verified for all pieces of equipment by computer simulation with the finite-element method combined with pilot tests (typically on a scale of 1:2) with all the handling equipment being submitted to dry-runs with a full scale concrete disk. This assured the functionality of the different pieces of equipment.

Lifting, Transporting, and Setting Down. These handling operations are performed using a vacuum lifting device (suction lifter) in conjunction with a crane in the manufacturing hall to lift the mirror blanks from the different support systems, to transport them within the building to different locations, and set them onto different support systems (mold for casting and coarse annealing, support for ceramization and thermal post-treatment, turning device, grinding machine, support for quality inspection, support for storage). The suction lifter must be universally equipped to handle blanks with different geometries and surface conditions depending on the progress of the process.

The vacuum lifting device for the 8.2 m blanks (see Fig. 4.38) is equipped with 18 suction cups, 12 of which are located on an outer circle and the residual six on an inner circle. They are connected to the support structure by hydraulic cylinders so that the above concept is realized. The suction lifter is able to pick up the blanks at their convex side as well as at their concave side. The support structure also carries all components for the vacuum generation, for the hydraulic system, and the complete set of electrical controls.

A handling cycle using the suction lifter (see Fig. 4.39) consists of the following main steps:

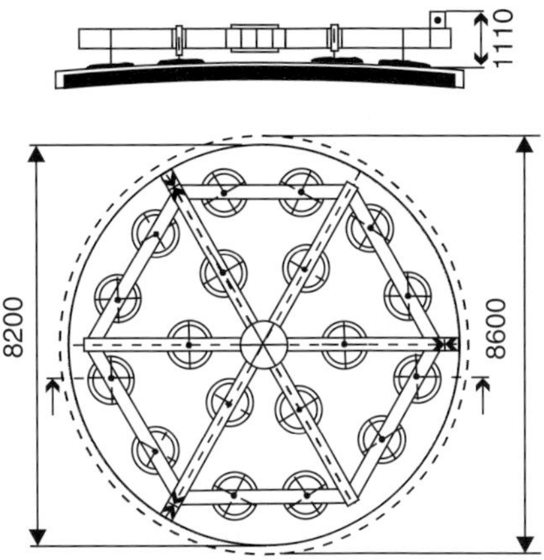

Fig. 4.38. Sketch of the vacuum lifting device for 8-m mirror substrates

Fig. 4.39. Vacuum lifting device during delivery of the first 8.2-m Zerodur® blank of 23 t

- functionality check of suction lifter and crane,
- preparation including functionality check of the support system onto which the blank will be set down after transport,
- positioning of the suction lifter over the blank using the crane,
- attaching of the suction cups to the blanks surface using the hydraulic system,
- evacuation of the suction cups,

- lifting of the blank from the support up to about 10 mm using the hydraulic system,
- lifting of the suction lifter including the blank to the height necessary for the transport, again using the crane,
- transport of the blank to the new position in the hall by driving the crane and its crab,
- lowering of the blank down to some cm above the new support,
- in case of turning device, grinding support, or quality control support, the single support elements of those are hydraulically brought into contact to the lower side of the blank,
- transmission of the blank load to the new support using the hydraulic system of the suction lifter,
- ventilation of the suction cups and withdrawal from the surface,
- storage of suction lifter using the crane.

There are two main aspects to be considered during the hydraulic lifting and lowering steps:

- as long as the blank is in contact with both the suction lifter and any support, the position of the blank should be defined by this support, and the hydraulic lifting or lowering, respectively, have to guarantee this in order to prevent the blank from being loaded undesirably,
- as long as the blank is in contact only with the suction lifter, the position of the blank must be defined by the latter.

The first aspect is met by hydraulically coupling all 18 cylinders of the suction lifter. This means that, on the one hand, the forces between the suction cups and blank are equal; on the other hand, however, the suction cups cannot define the position of the blank. The second aspect is satisfied by grouping the hydraulic cylinders in six arranged in a 120° sector of the blank each and by controlling the three groups separately. The transition from one condition to the other in the hydraulic system of the suction lifter is performed when the support is loaded only with a small part of the blank weight; this requires a very accurate control of the blank's position relative to the support and very small lifting or lowering steps performed by the three hydraulic groups.

Figure 4.40 demonstrates the global deformations of a mirror blank which is carried on the vacuum lifting device. These deformations will be approximately 200 μm with the maximum of the tensile stress at the surface being about $0.9\,\mathrm{N/mm^2}$.

Turning. The mirror blanks must be turned several times so that they can be machined on both sides. The turning device constructed for this purpose is also able to carry blanks of all geometries between the raw size and the final state.

The turning device (see Fig. 4.41) first consists of two cover parts, similar to the vacuum lifting device, and the blank is placed between them. The

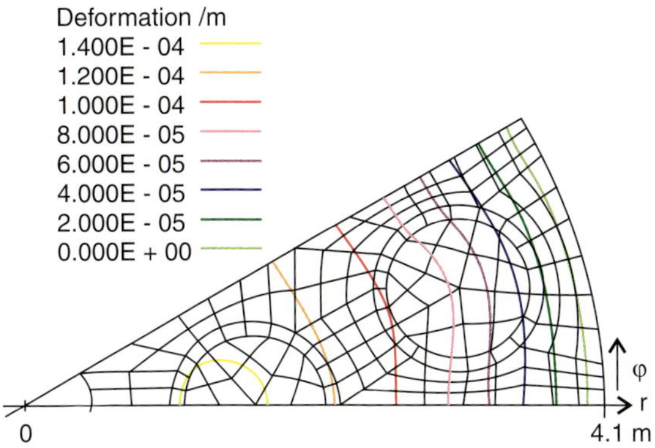

Fig. 4.40. Lines of equal deformation of the mirror surface in the axial direction relative to the zero position (0), for a Zerodur® mirror of 8.2 m in diameter and 175 mm in thickness, suspended from the vacuum lifting device; maximum deformation is 150 μm

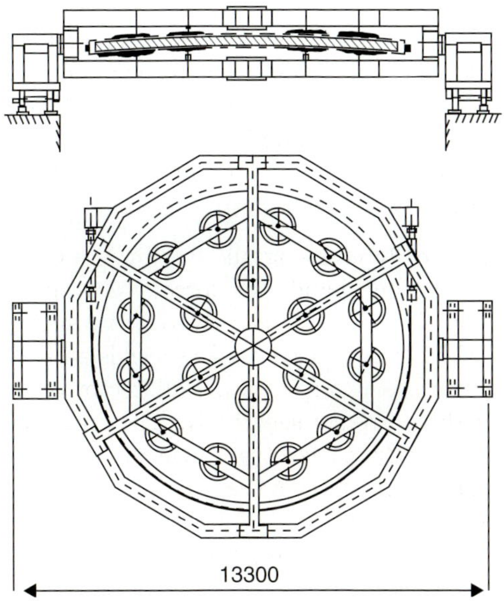

Fig. 4.41. Sketch of the turning device for 8-m mirror substrates

cover parts, having 18 support elements each, are connected by a ring-shaped frame which is mounted on two bearings to provide for turning of the whole arrangement. A special belt carries the mirror weight in the lateral direction, especially in the vertical position of the blank. It works at the lower half of the blank circumference and is mounted to the ring-shaped frame. The drives of the 2 × 18 support elements, the special belt for the lateral support, and the turning itself are hydraulic ones. The complete set of electrical controls are placed in a separate cabin.

A turning cycle (see Fig. 4.42) consists of the following main steps:

- functionality check of turning device;
- removal of the top cover from the turning frame using the crane;
- lowering of the blank down to some cm above the 18 support elements of the lower cover, using the suction lifter and the crane;
- bringing the 18 support elements of the lower cover hydraulically into contact to the lower side of the blank;
- transmitting of the blank load to the lower cover using the hydraulic system of the suction lifter;
- removal of suction cups from the blank and suction lifter from the turning place;
- attaching of the top cover to the turning frame and bringing its 18 support elements into contact with the top side of the blank;
- adjusting of the blank position in height so that the center of gravity is on the turning axis, using the hydraulic system of the lower cover;
- attaching of the special belt to the circumference of the blank (due to elastic deformation of the special belt when it is being increasingly loaded during the turning process the blank moves laterally about 20 mm; therefore, the blank is moved 10 mm in the opposite direction by prestressing the special belt);

Fig. 4.42. Turning device in action at about 45° turning angle

- hydraulic lowering of the blank by 15–25 mm (depending on the actual weight) taking into consideration that the blank moves in axial direction by double as much due to elastic deformation of the two covers and the frame during the turning process;
- opening of the ceiling of the machining cabin situated under the turning device;
- hydraulic turning by 180° with intermediate stops for checking;
- closing of the ceiling;
- removal of the top cover (lower cover before turning);
- removal of the special belt from the circumference of the blank;
- lifting of the blank using the suction lifter.

Finite-element calculations were also performed for the blank being placed in the turning device in order to investigate stresses and deformations. The maximum value of tensile stresses at the surface are $< 0.3\,\mathrm{N/mm^2}$ for the horizontal position, whereas they are $< 0.8\,\mathrm{N/mm^2}$ for the vertical one. Figure 4.43 shows the deformations of the mirror in the horizontal position with the maximum value being $< 6\,\mu\mathrm{m}$.

Storage. Three storage supports serve as intermediate storage (buffer store) for blanks of various machining states. Nine large support areas arranged according to the whiffle-tree concept (balanced beams) are sufficient in this case, resulting in maximum tensile stress values of $< 1\,\mathrm{N/mm^2}$ and maximum deformations of $< 100\,\mu\mathrm{m}$.

The storage supports are constructed as insulated containers in order to prevent the blanks not already being ceramized from thermal shocks that may be induced during casting of a new blank or when the huge doors of the manufacturing hall are opened during wintertime.

The storage containers can be piled up – by use of the crane – in any given sequence at two places in the manufacturing hall where the foundations are

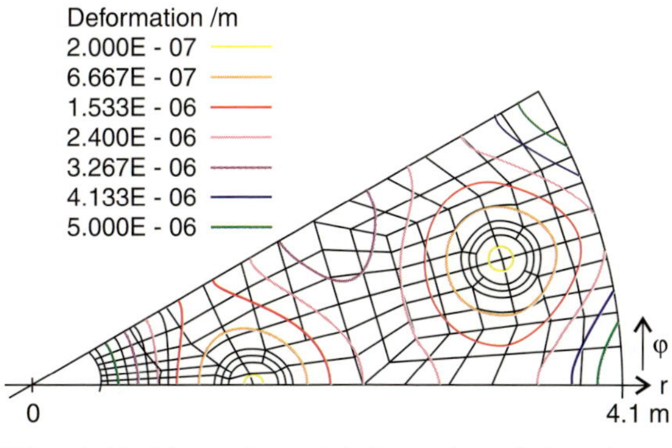

Fig. 4.43. Lines of equal deformation of the mirror surface in axial direction for a Zerodur® mirror of 8.2 m in diameter and 175 mm in thickness, supported by the turning device in the horizontal position; maximum deformation is 5.2 µm

constructed for a load of about 200 t each. When the blank is placed into a storage container the containers are set onto the floor. The transmission of the blank weight from the suction lifter to the storage support is done by the use of only the hydraulic system of the suction lifter. Subsequently, the container including the blank is piled up again at the normal storage place.

Machining

Owing to the large abrasion rate of today's diamond tools – with simultaneously good contour stability – easy handling and simple adaptation and change of the curvature radii, curved surfaces are mainly generated with bonded-grain tools. Loose-grain lapping is limited to precision finishing. In the past, it was customary practice in the manufacture of telescope mirrors to first polish the "best fitting sphere", as it is called, to test it by interferometer, and then produce an aspherical surface by lapping and polishing again. This procedure resulted from the relatively slight aspherical deviation of, for example, approximately 45 µm for a 3.6-m mirror (F/3.5). With today's short-focus mirrors, whose apertures range up to F/1, aspherical deformation is of the order of several hundred microns. Such aspheres can be machined economically and sufficiently rapid only with the aid of diamond tools.

The precision contours with a polishable surface finish, then, are made using the lapping process. Sufficiently high precision in diamond grinding, on the one hand, is a prerequisite (target value ±50 µm contour deviation) to come within the range of the interferometer used to determine the contours, which is required for precise lapping. On the other hand, efforts should be made to ensure that the microcrack depth caused during the grinding process will be within the amount abraded by lapping, which is ensured by the selection of suited diamond grain sizes. Given optimum machining conditions, grain sizes D 76–D 107 produce crack depths ranging from a maximum of 50 µm to 100 µm. The tools suited for this purpose are circumferential grinding wheels.

The precision of contours is not only determined by the machine and the tools, but to a great extent also by the workpiece fixture, i.e., the supporting system on the machine. Deformations suffered by the meniscus owing to its own weight in cases of insufficient support practically become a negative characteristic of permanent contour deviation upon the grinding process. This means that high portions attributable to deformation are abraded more heavily, which, upon release later on, leads to valleys and vice versa.

During machining, the 8-m blank is supported by 18 hydraulic support elements mounted to the turntable of the grinding machine. Stresses and deformations are thus limited in the turning device to those values that have been discussed for the horizontal position in the paragraphs on "Handling and Support". In order to achieve a sufficient accuracy of the contour, the hydraulic cylinders must be blocked during the machining operations.

Specifications were extremely tight. For all VLT mirror blanks, for example, the required exactness of the special convex zone of the supporting system was ≤ 20 μm. This value has been exactly achieved. For the concentricity of the center hole, a permissible deviation of < 1 mm was specified; the value actually achieved is 0.01 mm.

Performance

The quality of the mirror blanks was controlled in various inspection procedures. The CTE values were measured according to the sampling plan shown in Fig. 4.44. Sample rods with a length of 100 mm were cut from ceramized Zerodur® and measured with a high-precision dilatometer system. In Fig. 4.45 the results for mirror 2 are represented in a bar chart.

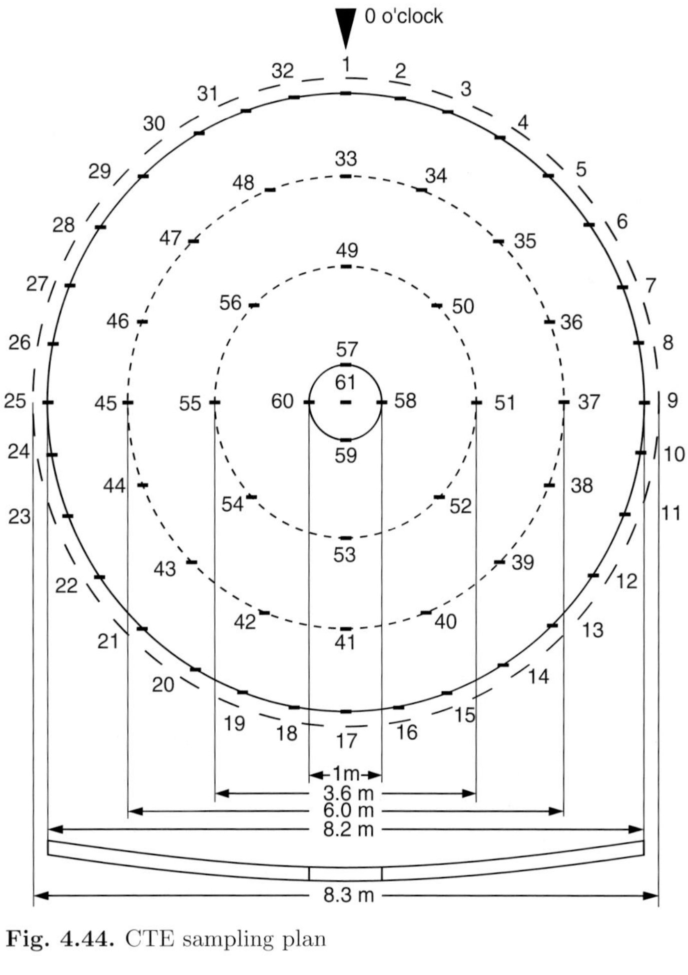

Fig. 4.44. CTE sampling plan

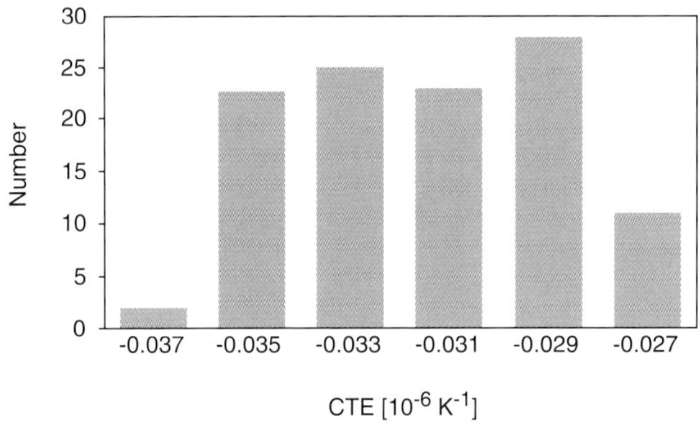

Fig. 4.45. CTE distribution measured for mirror 2 (0–50 °C)

The thickness measurements were carried out (normal to the convex surface) with an echo meter by determining the delay time of the ultrasonic signal reflected by the opposite disk surface.

The spherical surfaces were measured with different methods. First we used a theodolite system to measure the individual position of the *distance gauges* with the well-known triangular method. Then we measured 92 points of the VLT Zerodur mirror blank at 13 different diameters with sagittal height measurements (see Fig. 4.46). The special 300-mm diameter convex zones were measured with a nine-dial gauge device shown in Fig. 4.47. The test equipment consisted of a CNC grinder, three supporting points MP03, MP07, and MP11, nine dial gauges, and a calibration standard. 150 points at six different diameters were measured and evaluated. The resulting maximum peak-to-valley deviation from a perfect sphere was less than 20 microns.

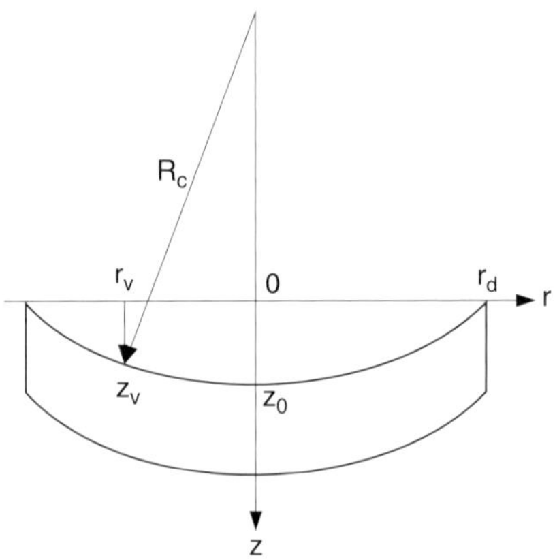

$z_v = \sqrt{R_c^2 - {r_v}^2} - \sqrt{R_c^2 - {r_d}^2}$
r_d: radius of mirror blank
r_v: radius
z_v: sagittal height at radius r_v
z_0: max. sagittal height at $r = 0$
R_c: radius of concave curvature

Fig. 4.46. Schematic of sagittal height measurements

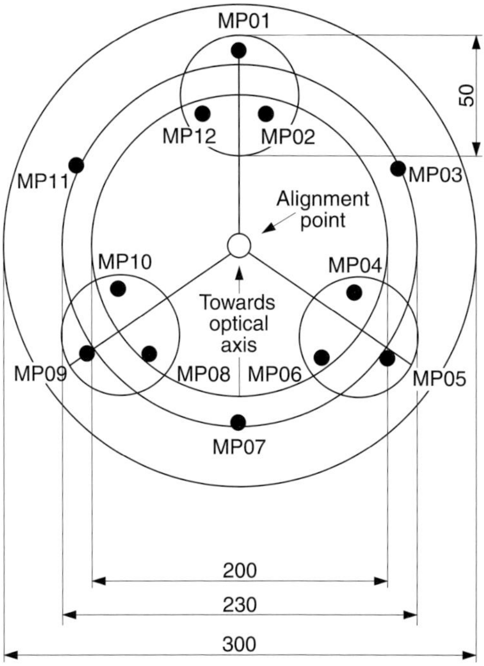

Fig. 4.47. Nine-dial gauge device

To control the internal quality of the VLT mirror blank, inclusions were inspected with a halogen lighting system and a long-distance microscope. Within an area of 25 mm below the concave surface (critical volume) we analyzed and evaluated

- the number of inclusions (bubbles, particles, etc.) with diameters > 0.1 mm,
- the mean and the maximum size of inclusions with diameters > 2 mm
- the average number of inclusions per cm^3,
- the maximum number of inclusions in any $10\,cm^3$.

The results are given in Table 4.2.

In order to measure the stress birefringences with the Senarmont method, the mirror blanks had to be placed on a quality control support system. The measuring points were distributed in such a way that the impact of residual stresses induced by the support was minimized. As an example, Table 4.2 lists the results of stress birefringence measurements carried out at a distance of 100 mm from the outer edge.

The permanent stress of the mirror blank was specified at the outer edge of the central hole and on all surfaces. The surface stresses were calculated with the help of finite-element models, which had been compared with the birefringence measurements.

Table 4.2. Quality features of the four VLT mirror blanks

Property/ characteristic	Specification	Results mirror 1	mirror 2	mirror 3	mirror 4
CTE for 0–50 °C (10^{-6} K^{-1})					
• mean value	(0.00 ± 0.15)	−0.043	−0.032	−0.040	−0.017
• homogeneity	< 0.05	0.009	0.011	0.024	0.028
Geometrical dimensions (mm)					
• diameter	8200 ±2	8201.52	8201.74	8201.72	8201.74
• thickness	176 +2/ − 0	177.9	177.7	177.5	177.7
Profile tolerance zone (mm)					
• concave surface	< 2	0.12	0.08	0.08	0.06
• convex surface	< 2	1.25	0.07	0.06	0.07
Radius of curvature; best fit (mm)					
• concave surface	$R_{\mathrm{concave}} = 28975$	28967.8	28970.1	28978.1	28976.3
• convex surface	$R_{\mathrm{convex}} = 28977$	28979.3	28979.9	28980.3	28979.9
Inclusions in the critical volume*					
• mean size (mm)	< 5	0.5	0.5	0.6	0.6
• maximum size (mm)	< 8	2.3	3.5	1.1	1.6
• average number (cm^{-3})	< 0.5	< 0.01	< 0.01	< 0.01	< 0.01
• maximum number per 10 cm^3	< 8	< 4	< 4	< 4	< 4
• maximum number up to 4 mm below concave surface	< 90	< 10	< 6	< 8	
• large number of inclusions in small volumes	none	none	none	none	none
Stress birefringence (nm/cm)**					
• mean value	≤ −10	−6.2	−9.3	−6.5	−8.0
• maximum value	≤ −20	−7.0	−10.4	−8.3	−9.5

* Two areas were defined for the mirror blank. The critical volume lies within 25 mm below the concave surface; the remaining volume was defined as "outside the critical volume".
** Negative values correspond to compressive stress.

As can be gathered from Table 4.2, the properties of the four 8.2-m Zerodur mirror blanks for the ESO/VLT not only fulfill but even surpass the extremely tight specifications and thus guarantee excellent performance.

4.3.2 Lightweight Mirrors

Lightweight Mirror Applications

One way to reduce the mass of a mirror is obviously to combine a thin meniscus with a complete active support system compensating for figure distortions under its own weight on a real-time basis. However, if the mirror is to be supported by a passive or only semi-active mount for cost, space, or weight reasons, lightweight mirrors are the logical solution. Among the various materials suitable for the production of lightweight mirrors, Zerodur® is the one with the highest thermal stability and the lowest CTE (see Table 4.3 for a comparison of the properties of several lightweight mirror materials).

One way to categorize applications for lightweight mirrors is to use the altitude of the system as a criterion. This leads to three different categories, each being characterized by specific cost tradeoffs.

Space applications obviously offer the highest cost savings if a lightweight mirror is used instead of a solid substrate when considering the cost of launch at several $10\,000\,\mathrm{kg}^{-1}$ (depending on the type of spacecraft and type of orbit for satellites). In addition, the feasibility and maintenance of a fully active or semi-active support is certainly much more complicated for a space-based mirror than for a ground-based or an airborne system. Examples of space-based systems with mirror substrates include weather, mapping, and surveillance satellites as well as optical and X-ray astronomical satellite telescopes and space probes.

In the case of airborne applications, the lightweighting cost has to be traded off against the flight cost of the weight saved over the lifetime of the system. Depending on the frequency, duration, and type of flight, this cumulative cost can be similar to the cost of a space launch. In addition, the weight-carrying capacity of the plane itself is obviously a limiting factor, too. Therefore, lightweight mirrors are used and considered for use in many

Table 4.3. Properties of lightweight mirror materials

	Zerodur	ULE	SiO_2	Boro-silicate	Be	SiC
Density ρ (g/cm)	2.53	2.21	2.20	2.2	1.85	3.05
Young's modulus E (GPa)	90.3	67.6	70	63	290	390
Specific stiffness E/ρ*** (10^{-6})	3.6	3.1	3.2	2.8	15.1	12.8
Thermal conductivity λ (W/mk)	1.46	1.31	1.38	1.12	159	185
CTE α (10^{-6}/K)	0.02 (0.01)*	0.03	0.55	3.2	11.4	2.6
Thermal stability α/λ** (10^{-8} mW)	1.37 (0.68)*	2.30	39.90	285.7	7.17	1.49

* on special request, ** small is best, *** high is best

airborne surveillance and telescope systems such as the SOFIA project (Solar Observatory for Infrared Astronomy).

Lightweight mirrors also offer significant advantages in various ground-based systems. Depending on the application, the costs for mounting a mirror increase more than linearly with the weight of the substrate. Potential cost savings for the mount can, therefore, be traded off against the costs of lightweighting the mirror blank. One of the most common examples is the secondary mirror of ground-based telescopes, especially if used in a chopping mode. Other examples of applications are a variety of components for precision instrumentation such as stages.

The cost savings offered by the applications will determine which method of lightweighting is applied in a specific case. While in certain applications weight savings of up to 70% achievable by machining may be sufficient, other applications may require lightweighting factors of 85% and more, which can be obtained with fused mirrors.

Mechanical Lightweighting

Monolithic, stiff pieces can be reduced in weight by 60–70% if holes are ground into them. The extent of lightening ultimately achievable depends on the specific geometry of the resulting light-weight structure, which is substantially determined by the rigidity required. Cavities are produced by grinding them out with the aid of diamond tools. One basically distinguishes between entirely open structures and undercut structures where, in order to boost rigidity, as high an amount of material as possible of the mirror rear side is retained. This latter procedure involves introducing the grinding tool into boreholes provided at the rear side. Both tool and shaft diameter determine the maximum size of the cavity, the tool being run – after introduction into the borehole – in a spiral fashion on an increasing circular path, until the tool shaft almost touches the wall of the cored hole (see Fig. 4.48).

Assuming a common shaft diameter of at least approximately 15 mm, the ratio of the cored hole to the undercut borehole is about 1:1.5.

By this process, it is possible to form not only round recesses but also any desired special shapes such as, e.g., triangles, rectangles and hexagons; owing to the tool diameter, however, the corners cannot be sharp-edged but only rounded; see Fig. 4.49. To achieve a residual weight as low as possible, it is desirable for the wall thicknesses of the structure to be largely reduced. The thing to do is to arrive at optimum values in the fields of high rigidity, saving of weight, and justifiable manufacturing costs; requirements which are highly difficult to meet at the same time. From experience, a 4 mm rib width and back plate thickness seems to be a reasonable value, while the face plate thickness is chosen to be larger by a factor of 2–3 because of print-through effects that may show upon polishing. Reducing the rib width and back plate thickness any further will boost manufacturing costs considerably, due to

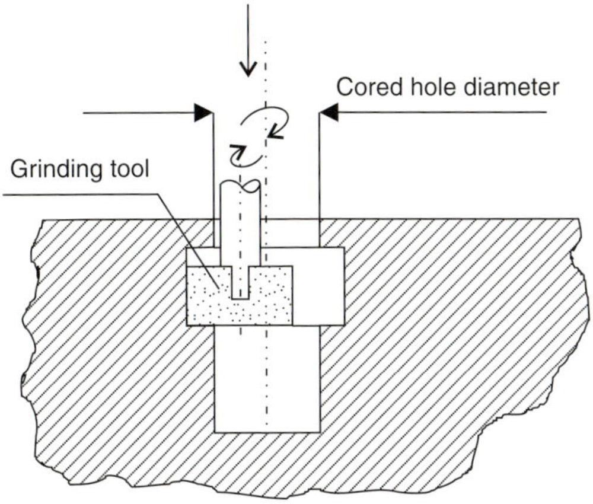

Fig. 4.48. Diagrammatic sketch of lightweighting by undercutting

markedly longer machining times and a high risk of fracture; the complete saving of the rear plate (open structure) is detrimental to the rigidity.

CNC Grinding. CNC grinding is a method of achieving weight reduction by machining hexagonal cells into the Zerodur® glass ceramic after blind holes have been drilled into the rear side of the massive blank. A lightweight factor of approximately 70% is achievable. By additional acid treatment, the factor

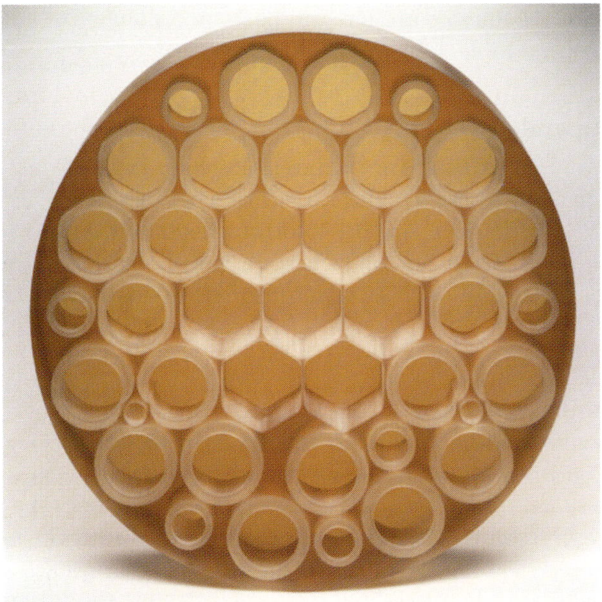

Fig. 4.49. Zerodur® lightweight mirror produced by grinding – study with different kinds of holes

can be increased to 80–85%. CNC grinding usually comprises the following steps:

- CAD testing of customer design and specification,
- design of grinding tools (shaft diameter, tool head with specific diamond grain size),
- programming of CNC-controlled precision machines, including test run on a PC,
- preparation of a quality-controlled Zerodur® block,
- testing of machine tool program and grinding tools by machining a typical section into a test block,
- drilling of blind holes into the mirror blank (including in-process inspections),
- machining of hexagonal cells and odd-shaped polygon cells (as transitions to the outer ring),
- detailed visual inspection after mechanical lightweighting,
- reannealing of the lightweight blank to reduce stresses and convert surface tensions into compressive stresses (or reduction of surface stresses by etching),
- handling, transport, and intermediate storage (use of special vacuum suction lifter, turning device, support system),
- issue of inspection certificate (including CTE measurements and further results according to specification).

CNC-processed lightweight mirrors are integral parts in a number of advanced technology telescopes. Table 4.4 gives a short overview.

Fusion of Single Parts and Related Techniques

The fusion technique is used for the fabrication of Zerodur® lightweighted mirrors with significantly more than 60% weight reduction. One possibility is to generate the rib structure by welding prefabricated thin plates with subsequent fusion to a front plate and a back plate, if required. Figure 4.50 shows a corresponding example. In the case of a curved front plate being required this curvature can be achieved by slumping (as described in Subsect. 4.3.1) a flat plate combined with the fusion in one process step.

A second option is to machine the rib structure from a massive disk by water-jet cutting, ultrasonic cutting, or conventional diamond tool machining for subsequent fusion to a front plate and possibly a back plate. Especially water-jet cutting is a very flexible process with respect to cavity shape because the "tool" – a high-pressure water jet mixed with an abrasive – is always the same.

Apart from the fabrication of lightweighted mirror substrates, the fusion technique is also well suited to fabricating other complex components, with

Table 4.4. Applications of Zerodur® mirror blanks lightweighted by CNC grinding

Earth-bound telescopes	Diameter mm	Weight kg	Weight per mirror area kg/m²	Weight reduction %
WIYN/Mirror 2	1200	115	102	70
WIYN/Mirror	776/1101*	59	88	66
MMT/Mirror 2	1714	290	125	70
Magellan/Mirror 2	1376	187	126	70
ESO/VLT/Mirror 3	880/1250*	108	125	66
Gemini/Mirror 2	1024	50	47	85
airborne:				
MSG-SEVIRI /Scan	530/830*	16	45	73
	* elliptic			

Space/airborne telescopes	Dimensions mm	Weight reduction %	Orbit/ duration	Launch
SPOT	Ø 735 × 85	70	820 km	1986
Meteosat	Ø 400 × 45	65	36 000 km	1978
Giotto	Ø 250 × 30		comet Halley	1986
Hipparcos astrometry	Ø 600 × 70	70	elliptic	1990
ROSAT	Ø 884 × 583 to Ø 470 × 583	8 conic tubes	580 km	1990
SILEX	Ø 280 × 400	special structure	LEO & GEO	1998
AXAF	Ø 1225 × 991 to Ø 632 × 991	12 conic tubes	shuttle orbit	1999
SOFIA	Ø 2705 × 350	75	14 km	2001

Abbreviations:
WIYN: WIYN Observatory owned by the WIYN Consortium (Wisconsin, Indiana, and Yale University and National Astronomy Observatories)
MMT: Multiple Mirror Telescope
ESO/VLT: European Southern Observatory/Very Large Telescope
SEVIRI: Spinning Enhanced Visible and Infra-Red Imager
SPOT: Solar Patrol Observing Telescope
ROSAT: Roentgen Satellite
SILEX: Semiconductor Laser Inter-satellite Link Experiment
AXAF: Advanced X-ray Astrophysics Facility
SOFIA: Stratospheric Observatory for Infrared Astronomy

cavities in particular, when Zerodur® is the material of choice but conventional processes cannot achieve the proper geometry. Weight reductions of more than 85% can be achieved by using the fusion technique.

The fusion technique for Zerodur® consists of the following important process steps:

- cutting of the single parts by sawing and grinding in the vitreous condition,
- polishing of the surfaces to be fused together,

Fig. 4.50. Zerodur® lightweight mirror sample of 485 mm in diameter and 12 mm in height, manufactured by fusing single ribs, faceplate, and backplate. The weight reduction is 80%

- etching of the polished surfaces,
- assembly of the single parts within an oven using a fixture that exerts a defined pressure on the surfaces to be fused together,
- heating to a temperature slightly above T_g; by this the fusion process takes place,
- unloading of the fused surfaces from pressure,
- cooling down to room temperature.

The temperature control can be performed either so that the fused structure remains in the vitreous condition or so that the ceramizing takes place immediately after the virtual fusion process. Zerodur® that is already ceramized cannot be fused, but fused structures being still vitreous can be subject to another fusion process.

During the fusion process – as well as during slumping; see Subsect. 4.3.1 – a slight nucleation and shrinkage take place due to the temperature; therefore, the time frame available for the process without premature crystallization is limited. The thickness of the single parts to be fused or slumped is limited to about 100 mm because the available time would not suffice for heating up greater thicknesses without the danger of breakage due to heat stress.

Direct Casting

Direct casting of lightweight mirrors means that the front plate is formed from melted glass in a single process together with the backside support structure (ribs). For this a certain amount of melted glass is poured into a mold followed immediately by the insertion of an arrangement of cores from above into the liquid glass. Thereby the glass rises between the cores, consequently forming the ribs of the support structure.

The greatest difficulty in the application of the direct casting technique is to find an adequate mold material which should have the following properties when coming into contact with the Zerodur® melt at a temperature of about 1400 °C:

- no chemical reactions (e.g., bubble formation),
- heterogeneous nucleation as slight as possible (surface crystallization is dangerous in respect to breakage due to the small rib thickness required),
- *either* no wetting by the glass melt and, therefore, no adherence between the mold and the cast piece (which is a presupposition for demolding),
- *or* wetting and adherence; the demolding, however, is easy because of the low strength of the mold material.

There are two groups of materials that harbor these requirements to a great extent: ceramic fibers and noble metal alloys. Figure 4.51 shows a lightweight mirror blank for whose manufacture ceramic fiber material was used. The direct casting of lightweight mirrors implies a great number of pieces of equal dimensions due to the high expense for the moulds.

4.3.3 Thin-Walled Cylinders

Zerodur® for X-Ray Astronomy

In 1895 *Röntgen* reported his discovery of X-rays, and in 1912 *von Laue* demonstrated with his famous experiments that X-rays are diffracted when passing through a crystal. It took another decade of research before *Compton* showed in 1923 that X-rays can be reflected from a polished surface. In recognition of the fact that the angle of incidence was so small with respect to the plane of the surface ($\leq 1°$) the term "glancing" or "grazing incidence reflection" was born. Obviously, total external reflection occurred which meant that the interaction of X-rays with matter could be described by an index of refraction, n, which was less than unity. For X-ray wavelength λ sufficiently far from absorption edges, n is calculated from (4.2)

$$n = 1 - \delta \,. \tag{4.2}$$

For a detailed description of the reflection at grazing incidence of X-rays, see e.g. [4.37]. X-ray astronomy began after World War II with rockets. The

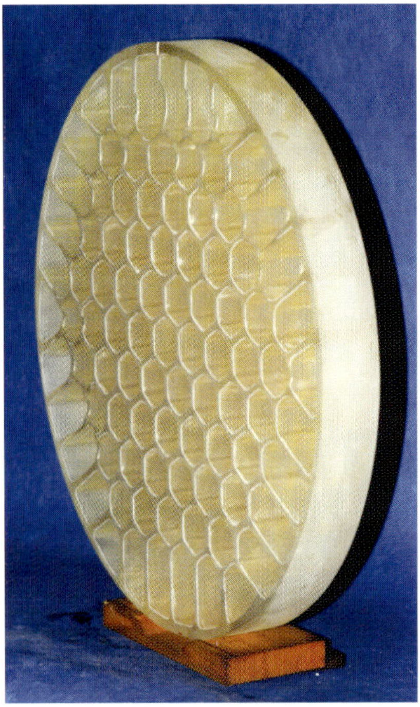

Fig. 4.51. Zerodur® lightweight mirror sample 664 mm in diameter and 72 mm in height, manufactured by direct casting by the use of a ceramic fiber mold. The weight reduction is 57% (faceplate thickness 12 mm, rib thickness 10 mm, cavity depth 60 mm, cavity width 50 mm)

first X-ray source, Scorpius X-1, was discovered more by accident during a rocket experiment conducted by NASA. These discoveries established X-ray astronomy and opened up the new research field of X-ray imaging optics.

In 1951 *Wolter* analyzed mirrors having concentric figures of revolution, i.e., paraboloids, hyperboloids, and ellipsoids [4.38]. He showed that in order to achieve a true image over an extended field of view the X-rays have to undergo two successive reflections from either a paraboloid/hyperboloid or paraboloid/ellipsoid combination which are mounted in a coaxial and confocal arrangement (see Fig. 4.52).

The index of refraction or the optical constants can be computed from dispersion theory. δ in (4.2) reads as follows:

$$\delta = N_0 \frac{Z r_e}{A 2\pi} \rho \lambda^2 , \tag{4.3}$$

where N_0 is Avogadro's number, r_e is the classic electron radius, Z and A are the atomic number and weight, respectively, and ρ is the mass density. In the case of heavy elements, for which Z/A is about 0.5, the critical angle for $\delta \ll 1$ can be estimated:

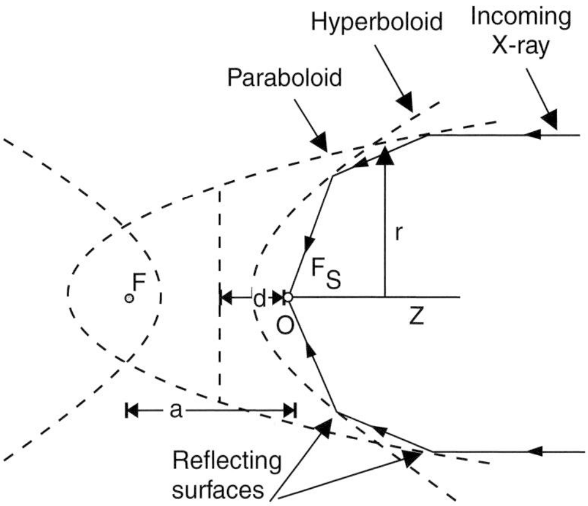

Fig. 4.52. Principle of the Wolter type I X-ray telescope: the telescope focus is at F_S, and the focus of the paraboloid is at the second focus of the hyperboloid. The convergences are considerably exaggerated and the surfaces actually resemble ordinary cylinders. The critical angle of incidence, usually less than 1°, depends on the mirror material, refractive index, and on the wavelength of the incident radiation. The paraboloid is in effect a "primary" mirror and the hyperboloid is a "secondary" that is added to shorten the optical path and correct for aberrations such as coma

$$\alpha_t = 5.6\lambda\sqrt{\rho}\,, \tag{4.4}$$

with α_t in arcmin, λ in Å, and ρ in $g\,cm^{-3}$. This means that for X-rays with a wavelength of a few Å the critical angle is about one degree. Equation (4.4) also shows the superiority of heavy elements like gold or platinum as X-ray reflectors. Therefore, efficient X-ray mirrors have to be coated. There are, however, exceptions to this rule if the heavy elements have absorption structures in the X-ray region of interest, such as gold near 5.6 Å. There the reflectivity is greatly reduced and lighter elements such as nickel become more favorable.

A polishing procedure had to be developed to achieve the desired paraboloid and hyperboloid form and to reach an extremely low level of surface roughness in order to provide a decent image. At a grazing angle of 2°, for example, a surface roughness with a height of about the X-ray wavelength used will scatter 20% of the total reflected X-rays off the image focus.

An example of a very successful X-ray telescope with Zerodur® mirrors is ROSAT. The resolution is 3 arc seconds. The X-ray detectors work in the energy range of 0.1–2.5 KeV (corresponding to 0.5–12.4 nm). ROSAT was sent into space in June 1990.

ROSAT (an abbreviation of "Roentgen satellite") is a German X-ray telescope having four nested tubular mirrors constructed with Zerodur® mirror substrates of 881 mm maximum diameter and 580 mm height. This system should have been sent up with the Space Shuttle in 1987. The loss of the

Challenger spacecraft destroyed these plans. After appropriate modifications it was finally transported into space by the DELTA-II rocket.

The AXAF telescope will have six concentric pairs of cylindrical mirrors (see Fig. 4.53). Including 12 spare blanks, 24 Zerodur® mirror blanks were ordered.

Zerodur® has outstanding properties for these applications: negligible thermal expansion, high Young's modulus and good geometric stability, very good working properties during grinding and polishing, as well as excellent chemical resistance which is very important during cleaning and during the processes of the vaporizing and dissolving of the reflecting metallic layers. The linear coefficient of thermal expansion α in the temperature range 0–50 °C is $0 \pm 0.05 \times 10^{-6}\,\mathrm{K}^{-1}$ ($\alpha_{\mathrm{steel}} = 11.5 \times 10^{-6}\,\mathrm{K}^{-1}$). For AXAF this coefficient was pushed to $0.08 \pm 0.05 \times 10^{-6}\,\mathrm{K}^{-1}$, adapted to the telescope system.

This is particularly important in telescope mirrors where a temporary change of shape due to temperature variation can lead to image distortion and hence to down time of the expensive equipment.

In Wolter telescopes, the negligible thermal expansion permits constant focusing of the short-wave electromagnetic X-ray radiation onto the detector. There are also considerable advantages when machining the material, since the heat generated in processing does not deform it, and hence the waiting time before interferometer measurements (which always accompany precision polishing) is minimal.

For the current ESA "X-ray Evolving Universe Spectroscopy (XEUS)" mission and for the NASA Constellation-X mission, components made from Zerodur® are being tested. The increasing effective areas of the telescopes require nested thin-walled mirrors of large diameters. The requested sizes of the mirror shells and the corresponding manufacturing tools almost exceed the reasonable technical limits of monolithic fabrication and handling. Mirror shells with diameters of more than 1 m are therefore manufactured

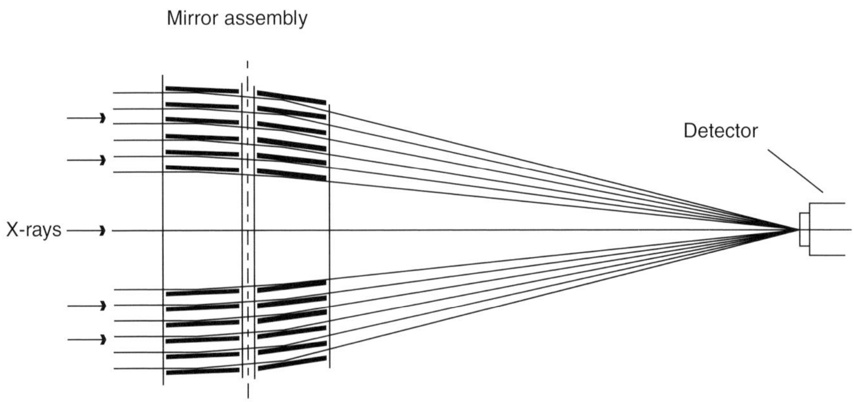

Fig. 4.53. Theoretical radiation of the AXAF telescope

in segments. For the mass production of segmented shells, the techniques of electroforming (XEUS) and epoxy replication (Constellation-X) are in evaluation. In both cases Zerodur® glass ceramic was chosen for the replication mandrels. The XEUS telescope (see Fig. 4.54) will contain 562 nested (Wolter type I) cylindrical mirrors with diameters between 1.3 m and 9.9 m, put together from several thousand mirror shell segments. Because each nested mirror has its own diameter, 562 different mandrels will be needed.

In 1999, Carl Zeiss was chosen by ESA to be the prime contractor of a demonstration mandrel for the XEUS project. The aim was to demonstrate the ability of polishing a segmented mandrel with a complex aspherical shape to the required accuracy. Schott was selected by Carl Zeiss to be the supplier of the pre-shaped mandrel. To meet the tight specifications, Schott controlled the quality of the mandrel in-situ with a 3D laser tracker system (measurement accuracy 10 μm). The profile tolerance of the aspherical shape was defined to be extremely narrow in order to reduce the subsequent lapping and polishing procedures carried out at Carl Zeiss with computer-controlled polishing (CCP) tools. The required parabolic and hyperbolic geometries were exactly prepared by CNC fine grinding. In a final quality control, Carl Zeiss measured the geometries and surfaces with high-resolution coordinate measuring machines, profilometers, and phase-contrast micro-interferometers. The first demonstration mandrel was delivered to the ESA in June 2000 (see Table 4.5).

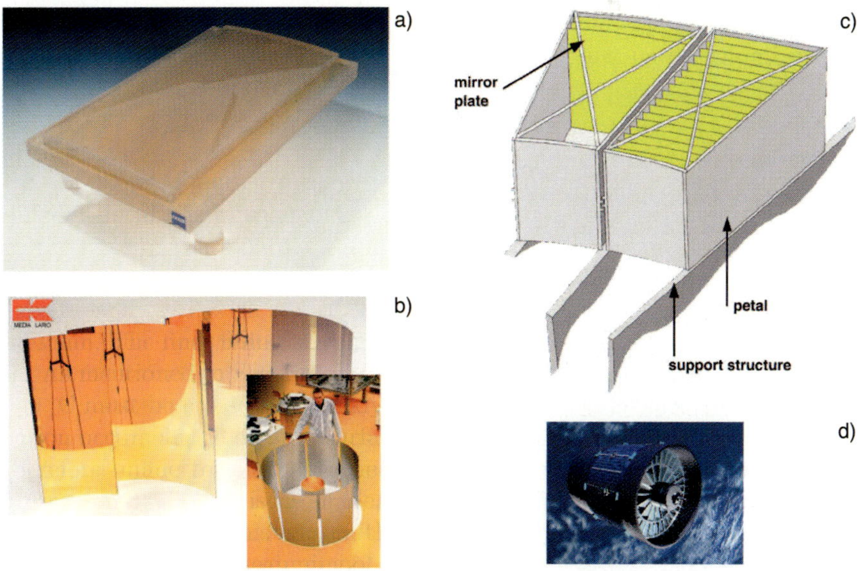

Fig. 4.54. Production of the XEUS X-ray telescope: (**a**) Zerodur® mandrel, (**b**) electro-formed mirror segments, (**c**) telescope segment, (**d**) model of XEUS

Table 4.5. Results of a Zerodur® mandrel prototype

Characteristic	Specification	Result
Focal length (mm)	50019.2	50014.5
Focal precision (mm)	± 20	−4.7
Cylinder radius (mm)	1900.729	1900.729
Surface microroughness (nm rms)	< 0.5	< 0.38
Resolution at 1 keV (arc sec)	2	2.4 (50% of foc. energy)

Manufacturing Process

Zerodur® is made in single melts each of which weighs about 40 t. Feeding in and melting of the raw materials lasts about two weeks. The melt is then homogenized, and by raising the temperature to over 1600 °C it is freed of gas bubbles (refined).

The temperature is reduced to about 1400 °C for casting, and the useful content of the crucible is cast into one or more preheated steel moulds. In a coarse annealing process, the temperature is reduced slowly to room temperature. In the case of the castings for AXAF this took about four months.

It is crucial to the whole of the manufacturing process that the melt remains sufficiently free of crystals so that it stays in a glassy state during machining and subsequent annealing. Later conversion into its partially crystalline condition is possible because of the difference between the temperature for maximum nucleation rate and the temperature for maximum crystal growth rate.

The maximum nucleation rate is achieved at a lower temperature (about 770 °C) than that for the maximum growth rate (about 1100 °C). Hence, the glass cannot crystallize during the initial annealing after casting in spite of the high crystal growth rate. This is because (except for a few heterogeneous nuclei), there are no crystal-forming nuclei present. By once again subjecting the glass to the temperature range for maximum nucleation, sufficient nuclei will be present for crystal growth during ceramization.

For the above reason, strict requirements are placed on the initial coarse annealing:

- Temperatures must monotonously fall during annealing down to room temperature. Local increases in temperature can lead to premature ceramization and breakage.
- Temporary stresses during annealing must be small. To prevent breakage during annealing, rapid loss of heat conductivity must be compensated for by a reduction in the cooling rate. Equally important is a good temperature distribution over the whole boule ($\leq 5\,°C$) in the furnace.
- Any permanent stresses present should be so small that the casting is capable of withstanding rough machining.

A first quality check and preparation for subsequent processing follows the coarse annealing. Then comes the all-important thermal process known as ceramization. By subjecting the glassy material to a temperature of about 800 °C it is converted into a partially crystalline state. This process is exothermic. For each kg of material, 250 kJ of heat are released. Furthermore, there is also a volume shrinkage of 3%. The temperature–time program chosen will determine both the crystallite size and distribution, and hence also the thermal expansion coefficient. In the case of the AXAF castings, the ceramization took eight months.

Before a casting is released for machining, all the inclusions in the entire volume must be located, catalogued, and measured. Because the crystals disperse visible light, transparency through a light path of 1000 mm is insufficient for normal internal inspection. For this reason infra-red light is used (see Fig. 4.55). The catalogue of inclusions enables the selection of the best mirror blank sizes to be obtained from each casting and the predetermination and optimization of the precise location from which they will be obtained.

During machining the stress symmetry is disturbed, and it must subsequently be re-established by fine annealing lasting about a month.

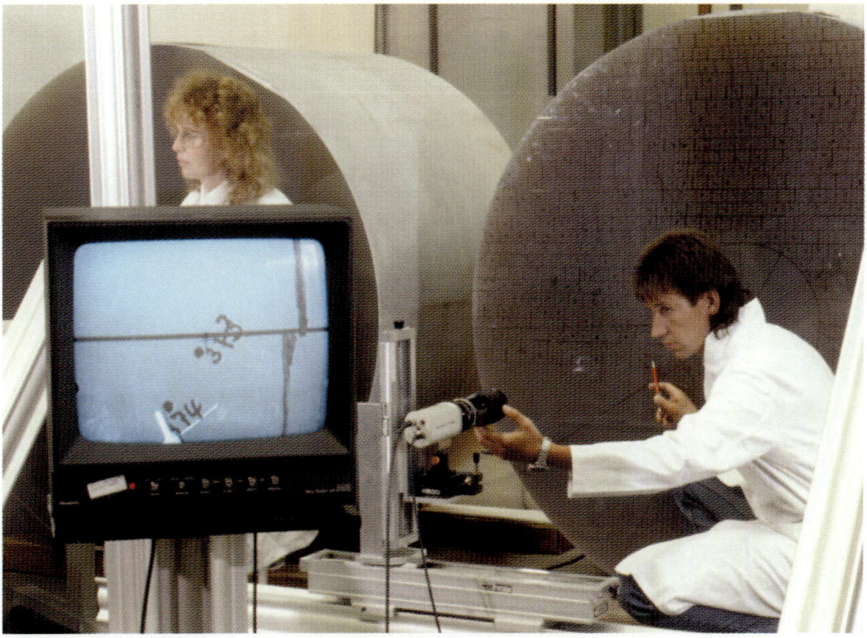

Fig. 4.55. Examining the raw castings for internal defects using an infra-red camera

Machining

Since thin-walled Zerodur® tubes cannot be manufactured directly by casting, they have to be worked out from solid cylinders. There are two aims attached to mechanical working. First, a maximum number of tubes have to be worked out from solid Zerodur® boules and, second, these tubes subsequently have to be ground to the final measures desired, with utmost precision and production safety being applied. Both aims can be achieved solely by using diamond tools on modern-type CNC machines. Separation of the individual tubes from the cylinder is performed by drilling-out (drill-grind process) with the aid of diamond core-drills (see Fig. 4.56). In practice, depths of bore of up to 1200 mm have been achieved already. The individual boreholes overlap slightly. As the telescope tubes are of conical shape, the boreholes have to be drilled, observing the same angle. The core is removed after drilling (see Fig. 4.57). In part, these rods are used as measuring samples for the purpose of testing the physical properties. After the outer tube is separated from the inner cylinder part, it is subjected to conical rough grinding and fine annealing, upon which the balanced stress condition disturbed by the drilling action is restored. The aim is to re-establish a condition where all the surfaces are under slight compressive stress. This is followed by the finishing process. Fixing the thin-walled Zerodur® tube safely and preventing it from being deformed by the clamping device constitute a specific problem in the observance of close tolerances. After the workpiece is detached from the machine, these deformations show up as circular runout. As a consequence, it is reason-

Fig. 4.56. Boring out the hollow cylinder with diamond tools

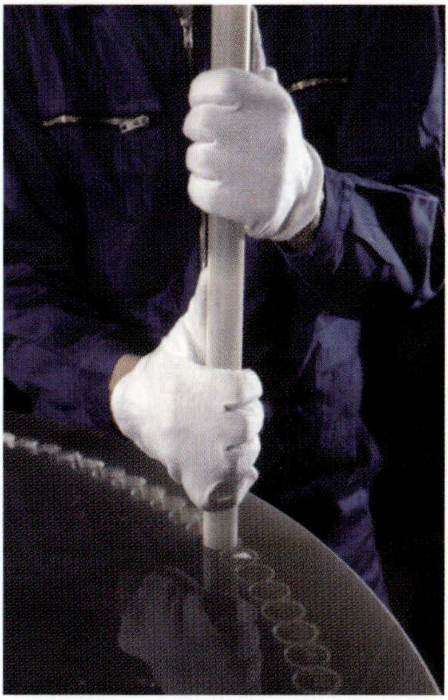

Fig. 4.57. Removing a core

able to first cause the workpiece, with the aid of air bearings, to be "floated in" free from stress on the machine. During the gripping action, continual measurements are needed to ensure there is no deformation. Of an equally critical nature are temperature effects – which Zerodur® of course does not react to – that may produce measurable deformations on both the clamping system and the machine and, thus, may ultimately cause the workpiece to vary in shape.

Schott supplied the mirror substrates for both X-ray telescope projects, namely ROSAT and AXAF. As far as production technology is concerned, the AXAF tubes, measuring up to 1274 mm in diameter and 991 mm in length, with wall thicknesses ranging from 17 to 24 mm, constituted the greatest challenge. Tolerances of 0.1 mm and axial runout of the front sides, as well as circular runout of the conical surfaces within 0.05 mm have had to be observed (see Fig. 4.58).

Handling

Transport and handling devices used for massive, i.e., stiffly stable workpieces, fail with thin-walled hollow cylinders. It has to be assured that handling forces do not affect the integrity of the workpiece, since unacceptable deformation can lead to fracture. The transport and handling device developed together with the industry was qualified to transport and to handle workpieces safely as well as to equip the cylinders on the CNC grinding machine.

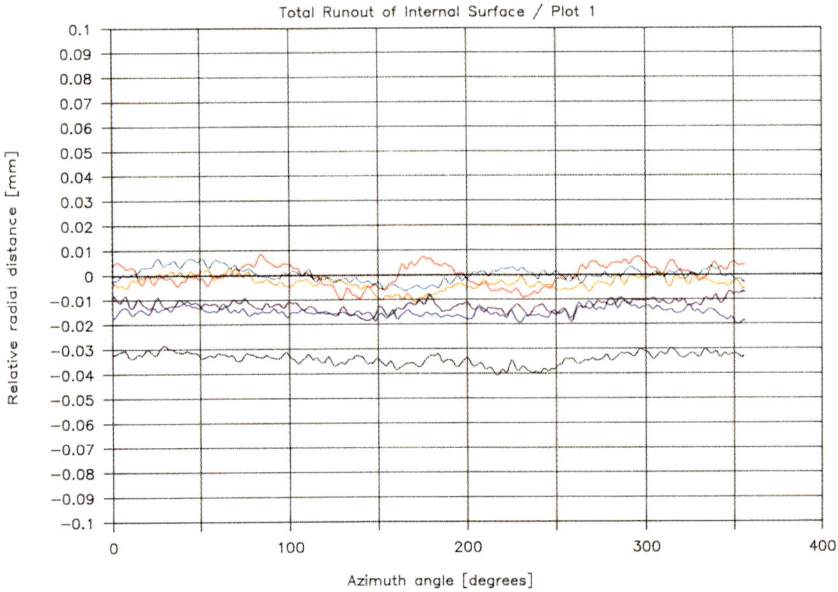

Fig. 4.58. AXAF mirror blank H2: total runout of internal surface

The handling device consists of a frame with integrated vacuum pump, a rail with pendulum arms, four vacuum suction cups, and a swivel motor (see Fig. 4.59). Whereas the pendulum arms can be adjusted to the required workpiece diameter, the suction cups have to be installed in order to fit the cylinder type to be handled. For slow lifting and lowering of the workpiece there is a hand-controlled hydraulic cylinder between the hard eye and the crane hook. Weight forces are controlled by digital crane scales.

The operating sequence for turning includes the following actions:

- installation of suction cups depending on the cylinder type,
- coupling the hydraulic unit with crane scales with the crane,
- attachment of handling device and hydraulic unit,
- check of crane and handling device (vacuum unit, suction cups, swivel motor, mechanics),
- marking of the suction cups' positions on the workpiece,
- preparation of CNC machine for handling,
- positioning with the crane,
- adjustment of pendulum arms,
- positioning with hydraulic unit,
- joining to the workpiece and evacuation of suction cups,
- lifting of hollow cylinder with hydraulic unit,
- lifting and transporting with crane above the turning point,
- turning by aid of the swivel motor,

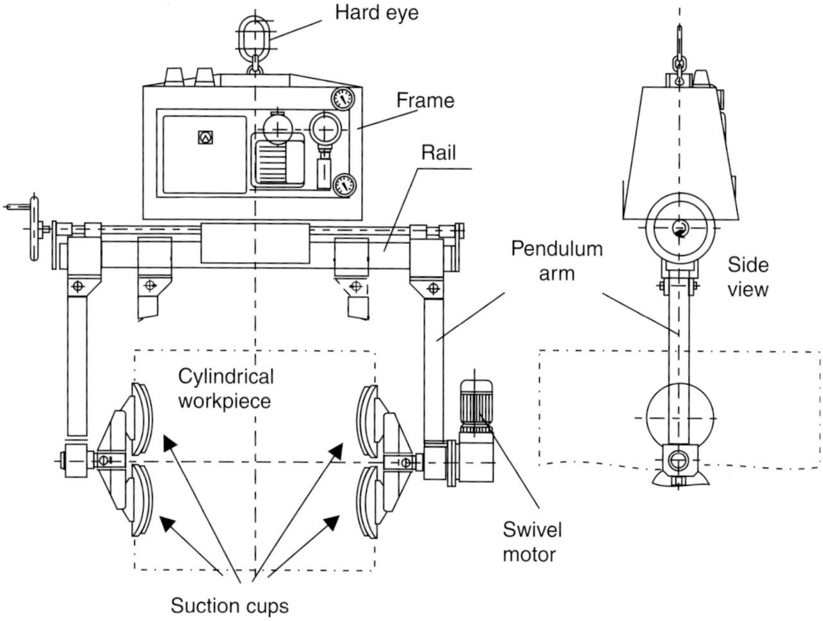

Fig. 4.59. Handling device for thin-walled hollow cylinders

- positioning above the machine table with the crane,
- lowering down to 30 mm above the machine table,
- smooth lowering down with hydraulic unit until the workpiece is completely lying down (weight comparison at crane scales),
- ventilation of suction cups,
- spreading of pendulum arms,
- moving of handling device and setting down with crane.

With this or similar procedures 24 thin-walled hollow cylinders for the AXAF project have been transported and handled successfully.

Quality Assurance

Inclusion Mapping of Large-Sized Boules. The casting procedure of Zerodur® boules with large dimensions (diameter 1.5 m × 1.2 m) may be disadvantageous in terms of the internal quality: cooling down must be very slow, so crystal growth can lead to macroscopic crystal inclusions of a few mm in diameter.

Specification requirements for AXAF mirror blanks ask for inclusion sizes smaller than 0.8 mm in diameter for the critical volume, which is the zone 5 mm below the internal surface.

Boules thicker than 600 mm are not accessible to visual inclusion inspection due to the high transmission loss. Inspection with a highly sensitive low-noise CCD camera, however, is still possible in the near infrared spectrum ($\lambda = 800$–1000 nm) (see Fig. 4.55).

By scanning the 1.5 m diameter boule surface with a field of view of 50 mm, the detection of inclusions down to 0.3 mm is possible in the whole volume. Focusing in the z direction provides the axial position of the inclusion, radial and azimuthal positions are obtained by projection of the inclusion's position to the polished front surface of the boule. After mapping of the entire boule, a computer optimization procedure allows us to find the best position of the final mirror blank in the boule.

Since AXAF mirror blank diameters range from 0.6 to 1.3 m, the optimization for AXAF boules allowed the extraction of up to three blanks from one single boule with inclusion contents according the specifications.

Metrology of Net-Shaped AXAF Mirror Blanks. Grinding of Zerodur® tubes of 0.9 m diameter and 20 mm wall thickness, which represent median values for AXAF mirror blanks, is a challenge in terms of elastic deformation. The low tolerances for AXAF mirror blanks (e.g., 0.15 mm for total diameter tolerance, 0.10 mm for surface runout) require not only special efforts for the grinding procedure, but also a special metrology adapted to the boundary conditions in the grinding shop.

The metrology for AXAF mirror blanks was designed for measurement on the grinding CNC machine. Diameter and surface runout measurements were made by a contactless triangulation sensor. Accuracies of 0.03 mm for the diameter and surface runout measurements were obtained. The in-situ measurements on the grinding machine, without removal of the mirror blank from the machine table, were feasible due to calibration of the machine coordinates with Zerodur® low-expansion rods.

4.4 Optical Figuring of High-Quality Optical Surfaces

Ernst-Dieter Knohl

4.4.1 Introduction

With Zerodur® blanks, Carl Zeiss is able to finish optical surfaces of outstanding quality because of the homogeneity of the material, the low thermal expansion coefficient, and the inert behavior of the material during the polishing process.

The best surface roughness ever made has been achieved for the ROSAT (Röntgen satellite) mirrors with about two ÅRMS (root mean square) microroughness. This value has been measured by X-ray scatter measurement at the PANTER test facility of the MPE (Max Planck Institut für Extraterrestrische Physik). The contour accuracy of the mirror shells with parabolic shape in the primaries and hyperbolic shape in the secondaries reached an

X-ray performance of three arc seconds HEW (half energy width) for the complete mirror system of Wolter type I with four concentric nested mirrors [4.39].

ROSAT was launched on June 1, 1990, and is now collecting excellent sharp images of an expected 100 000 new X-ray sources in the sky [4.40, 41]. Another example of superb optical surfaces with Zerodur® mirrors is the optical system for the ESO (European Southern Observatory) New Technology Telescope (NTT) with a 3.6 m primary mirror. Primary, secondary, and tertiary mirrors have been finished with extraordinarily smooth surfaces, and in connection with the active correction of the primary, this telescope marked a milestone in modern astronomy.

The primary can concentrate 80% of the light energy to better than 0.1 arc seconds diameter. That means that faint objects of 29 mag (magnitudo) can still be resolved within a reasonable exposure time [4.42].

In comparison to earlier 4 m class telescopes the NTT primary has a much shorter focal ratio of F/2.2 and the resulting aspherical deviation from the best-fit sphere is in the range of 200 µm. The aspect ratio was chosen to be $D/h = 3600/240 = 15$ and allows active correction of a few λ by the 78 actuators of the axial support system. Because long-wave errors such as astigmatism or spherical aberration can be corrected, the demand for an extremely smooth optical surface becomes much more stringent [4.43]. This goal has been reached with:

- calibrated mirror support with air bearings to release all restraints during testing,
- real-time interferometry (Direct 100),
- membrane technology.

Since June 1989 the ESO NTT has been in operation at La Silla in Chile and the astronomical community has been fighting for observation time.

4.4.2 Lapping of Aspherical Surfaces

Starting with a preground shape of the Zerodur® blank, optical figuring in general is done with loose abrasive and with the help of tools that are able to find the workpiece. Thus, the requirements for machining accuracy can be relaxed. The lapping process is about one magnitude slower than grinding.

To reach a large wear rate, a coarse grain size and large tools are used. The last point is of essential importance to avoid ripple structures because they become visible in only the final polishing phase. The tool size and stroke amplitude define the final surface smoothness with respect to short-wave errors.

A fast metrology is needed to control the lapping process and continuously to obtain the exact topography of the optical surface. With that information the tool shape can be adapted to the next figuring step, as can the function of the tool wear rate. Lapping and testing become an iterative process.

To obtain the surface information during lapping, at Carl Zeiss one has used an IR interferometer with stabilized CO_2 laser and $\lambda_{IR} = 10.5\,\mu m$ wavelength. With a resolution of $\lambda/20$, about 500 nm surface quality can be inspected in the lapping phase. This technique allows measurements of non-reflecting surfaces. To overcome the wavefront deviation of aspherical surfaces, IR null systems are used for wavefront correction. IR interferometry is a fast tool for inspecting mat surfaces if local gradients do not exceed the dynamic range of the sensing pyroelectric Vidicon. One disadvantage of the null correction is that the system is working correctly for only small aberrations. In case of large aberrations – as usual – there is local distortion and one has to correlate the fringes to the topography.

The null system can either be made as a reflecting system or as a lens design. Typical IR materials for lenses are germanium, barium fluoride, or zinc selenide.

In modern telescopes the focal ratio has been further reduced to typically F/1.0, i.e., the aspherical deformation reaches values of a few millimeters. Then the figuring process can no longer be started with a spherical surface and a high-quality ground asphere is required to begin with. As a consequence the grinder has to work with an accuracy of a few µm, the grinding process has to be controlled exactly, and a high-quality system is needed to support the blank.

4.4.3 Polishing of Aspherical Surfaces

In the fine-smoothing phase the grain size of the slurry becomes finer and finer down to 1 µm diameter and has to be absolutely uniform. The tool pressure and viscosity of the slurry have to be chosen so that "aquaplaning" is avoided. The wear rate is one to two magnitudes smaller than during lapping.

The surface becomes shiny and is tested with a visual interferometer. A He-Ne laser with 632.8 nm wavelength is used. Then the fine structure of the surface becomes visible. In the worst case one has to return to the lapping process if ripple amplitudes are too large to be reduced by polishing.

With a Vis-interferometer testing becomes very sensitive so that vibrations and local seeing influence the fringe pattern. Then, the testing equipment can be decoupled from vibrations up to 1 kHz by the use of real-time interferometry with the Direct 100 system and a 3D stabilization. Now interferometric tests are possible at any time with a resolution of 1 nm or even better.

In case of concave surfaces the aspherical correction of the wavefront is done with a null system. If a convex surface has to be tested, such as the surface of a secondary mirror, testing is performed against a concave matrix glass.

For large mirrors with fast focal ratio, the membrane tool technology has been developed. Polishing is done with a strip-like membrane tool to correct the radial profile, and with a set of pressure-controlled tools the non-rotation-symmetric errors can be worked on. The membrane acts as a low-pass filter, avoids ripple structures, and is always in contact with the whole mirror surface.

If the wavefront aberrations become smaller than 100 nm RMS, the surface gradients have to be analyzed and the encircled energy of the light is evaluated.

The influence of different error types is studied in a mathematical model simulation. The best surface quality in the final smoothing phase is obtained by such an optimization.

4.5 Special Characteristics of Zerodur®

Reiner Haug, Wilfried Heimerl, Burkhard Speit

4.5.1 Length Stability

Introduction

Zerodur® as a homogeneous, non-porous glass ceramic exhibits excellent dimensional stability not only during long-term isothermal applications but also when it is subjected to temperature changes within a broad range of applications. Because of this outstanding feature Zerodur® has been used for many years in a permanently widening field where the greatest length stability is required.

Originally developed as a mirror blank for large astronomical telescopes, Zerodur® is today also used in X-ray telescopes and in a number of high-precision optical applications, such as laser gyroscopes, reflective optics for microlithography, or distance-defining parts in laser resonators.

For all these applications Zerodur® has to be used because its coefficient of thermal expansion (CTE) is practically zero within a large temperature range. This very low CTE is achieved by the special microstructure of the glass ceramic Zerodur® which comprises two phases:

- a crystalline phase consisting of h-quartz solid solution crystals with an average size of 50 nm and a negative CTE,
- a remaining glassy phase with a positive CTE.

Both phases can be adjusted in such a way that the overall CTE nearly vanishes.

As the requirements for the stability of high-precision instruments in which Zerodur® has to be used because of its benefits are increasing progressively, it has become apparent that there is a limit to dimensional stability: if the relative length changes $\Delta l/l$ of a Zerodur® part are required to remain of the order of 10^{-5} or less, certain temperature cycles such as heating up to 300 °C and then quenching to ambient temperature are not allowed. In other words: if very fine relative length changes of the order of $\leq 10^{-5}$ are of interest, it must be realized that the length of Zerodur® is not only a function of temperature but also of cycling rates, i.e., of the thermal history. *Bennet* [4.44] was the first to publish these effects and a number of corresponding publications have since followed [4.45–48].

Measurement of the Length Stability

The CTE of Zerodur® is influenced during three stages of the production process:

- the batch composition of the melt – influences the composition of the h-quartz s.s. crystals;
- the ceramization process – determines the crystal contents in the glass ceramic;
- the annealing process (see below).

After each of these three independent stages the CTE value of the later Zerodur® item has to be measured and the results have to be used as feedback to supervise the process. The CTE measurement therefore has to be taken rapidly (as results must be available within hours), so it must be easy to perform and exact to a maximum level of accuracy.

As a standard characterization the mean CTE (0/50) value in the temperature interval between 0 and 50 °C is measured in pushrod dilatometers made of titanium silicate (ULE® 7971). An inductive transducer (LVDT) registers the motion of the pushrod with a resolution of 5 nm. The electromechanical sensitivity of the LVDT is calibrated with a Michelson interferometer. The expansivity of the dilatometer itself is determined by measuring standard materials for calibration. The standard materials (ULE®, fused silica) have been measured especially for Schott in vacuum interferometers by the National Standardization Institutes PTB (Germany) and NPL (GB).

The sample is a rod with a length of 100 mm and a diameter of about 5 mm and with ground and polished ends. The tempering of the samples during measurement takes place in water or in an air–helium atmosphere depending on the temperature range.

The dilatometers show an uncertainty (95% confidence level) of the measured relative length change of the sample of

$$u_{\Delta l/l} = \pm 0.4 \times 10^{-6} \quad (\pm 0.25 \times 10^{-6} \text{ in case of water tempered dil.}) \,,$$

and thereby allow measurements of CTE and observations of length stability in the temperature range from −100 °C up to +300 °C to a very high degree of precision. For example, the CTE (0/50) of Zerodur® is measured with an uncertainty of

$\pm 5 \times 10^{-9}\ \mathrm{K}^{-1}$.

Interferometric measurements lead to even higher degrees of accuracy; on the other hand, these measurements are distinguished by very expensive sample preparation procedures and very long measuring times. Interferometrical measurements were performed with Zerodur®, investigating its length stability by the above-mentioned standardization institutes [4.44, 45, 49, 50].

Thermal Cycling

There are three temperature ranges where length stability may be diminished by changes of expansion or CTE:

(a) $T > 700\,°\mathrm{C}$ (ceramization range)
(b) $130\,°\mathrm{C} \leq T \leq 320\,°\mathrm{C}$ (upper relaxation range)
(c) $-70\,°\mathrm{C} \leq T \leq +40\,°\mathrm{C}$ (lower relaxation range)

Ceramization Range. At about 700 °C the ceramization continues (crystal growth) and leads to irreversible changes of the dimensions and the material properties; the CTE in all temperature ranges is changed irreversibly (see Fig. 4.7 in Sect. 4.1).

Upper Relaxation Range. Figures 4.60, 4.61, and 4.9 (see Sect. 4.1) show the limits and the influence of this temperature range on the dimensional stability. Initially well-annealed samples were quenched from different temperatures. Figure 4.60 shows that until quenching temperatures of 130 °C no change of the CTE (0/50) happens; with increasing temperature the CTE (0/50) increases linearly until at 320 °C a saturation value is achieved. A total rise of the CTE (0/50) from $-50 \times 10^{-9}\,\mathrm{K}^{-1}$ to $+20 \times 10^{-9}\,\mathrm{K}^{-1}$ is observed. At temperatures above 320 °C no further variation of the CTE occurs.

Figure 4.61 shows the influence of the cooling rate upon the CTE (0/50): the CTE is proportional to the logarithm of the cooling rate.

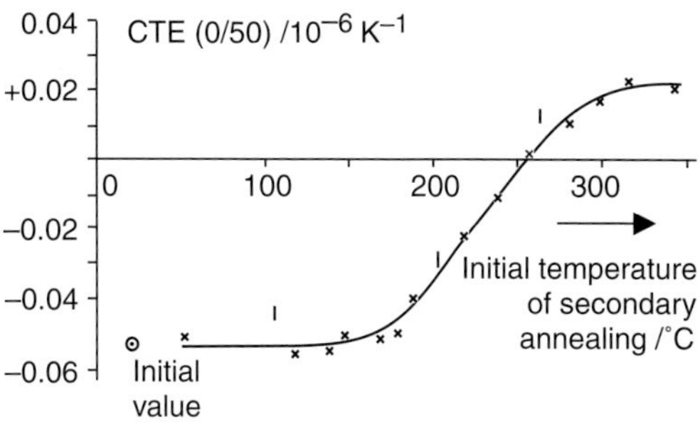

Fig. 4.60. Variation of the coefficient of thermal expansion CTE (0/50) (0–50 °C) of Zerodur® – first cooling at 0.1 K/min – as a function of the initial temperature of a secondary cooling in open air to room temperature

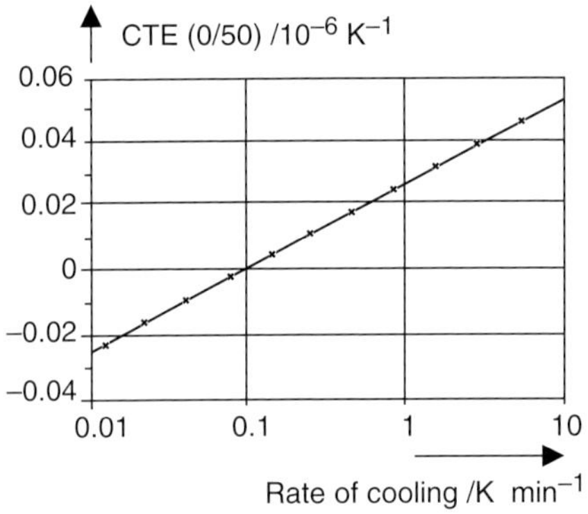

Fig. 4.61. Variation of the coefficient of thermal expansion CTE (0/50) as a function of the cooling rate from 320 °C to room temperature

Thermal cycles up to 300 °C and down again with different rates lead to dimensional changes as Fig. 4.9 shows. Fast cooling leads to a greater length (lower density). The 30 K/h cooling curve leads to a relative elongation of 14×10^{-6} for a sample which has originally been annealed with a rate of 1 K/h. If this sample had been quenched from 300 °C, an elongation of 20×10^{-6} would have been obtained.

All these effects are strictly reversible (as long as the temperature is kept below 700 °C) so that a sample, whatever temperature history it has been subjected to in the meantime, will return to its original dimensions and CTE as soon as it is heated up to above 320 °C and then annealed with its original rates. On the one hand, this phenomenon provides a very powerful way to adjust the CTE of Zerodur®. On the other hand, it might present a disadvantage for users who must subject Zerodur® parts to the upper relaxation range during a production process. This could force users to re-anneal the Zerodur® item again from 320 °C with the original rates. The effects in the upper relaxation range are caused by the MgO content of Zerodur® [4.47]. Schott has, therefore, developed the MgO-free Zerodur M® (see below) for which the above-described effects are not measurable.

Lower Relaxation Range. Thermal cycles which touch or include the lower relaxation range lead to a hysteresis shape of the $\Delta l/l\ (T)$ curve (see Fig. 4.62). The vertical opening of this hysteresis is of the order of $\Delta l/l = 10^{-6}$.

Like the effects of the upper relaxation range these effects are also strictly reversible. The effects of the upper relaxation range, however, are stable in the room temperature range, while those in the lower range relax in a relatively short time even at rather low temperatures (see Fig. 4.63).

Isothermal Long-Term Stability

To study the long-term stability, Zerodur® bars with a length of 400 mm were observed at a constant temperature of 20 °C at the PTB in Braunschweig

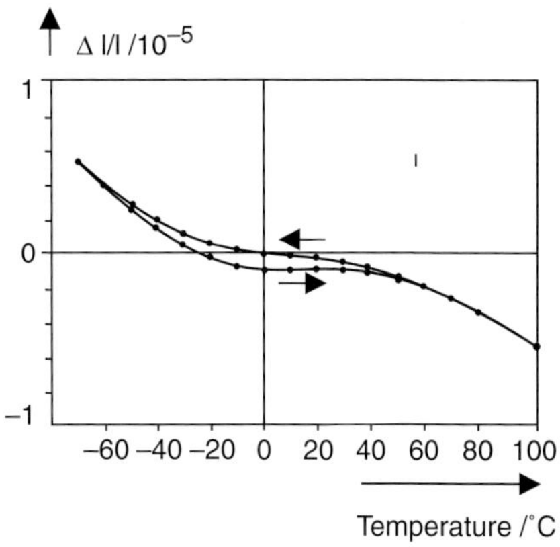

Fig. 4.62. Hysteresis in the course of relative length change $\Delta l/l$ during a thermal cycling process of Zerodur®, temperature rate = 1.25 K/min

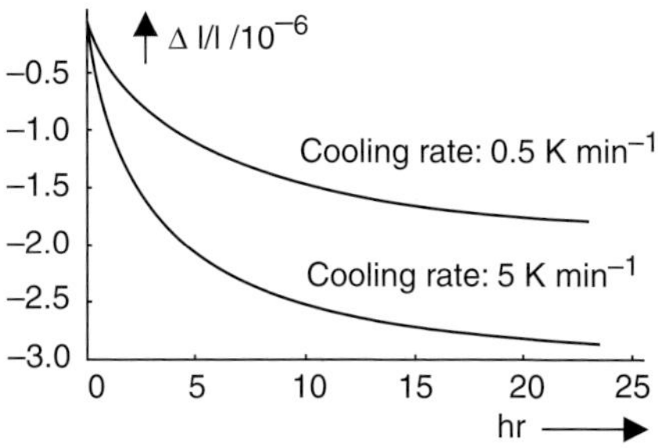

Fig. 4.63. Isothermal change of relative length of Zerodur® at −30 °C after cooling from +20 °C to −30 °C at 0.5 and 5 K/min

from September 1973 (production of the material in 1971, last annealing in May, 1973) until 1983 [4.50].

The length measurements were performed interferometrically with a Kösters Komparator so that the measure was realized by the wavelength definition of the Meter of 1960. (Any material measure was excluded in that special case – which material should act as measurement standard for measuring Zerodur®?!)

A typical result is shown in Fig. 4.64. The shrinkage velocity is identified by the longitudinal ageing coefficient A in the formula $\Delta l/l = A \cdot \Delta t$,

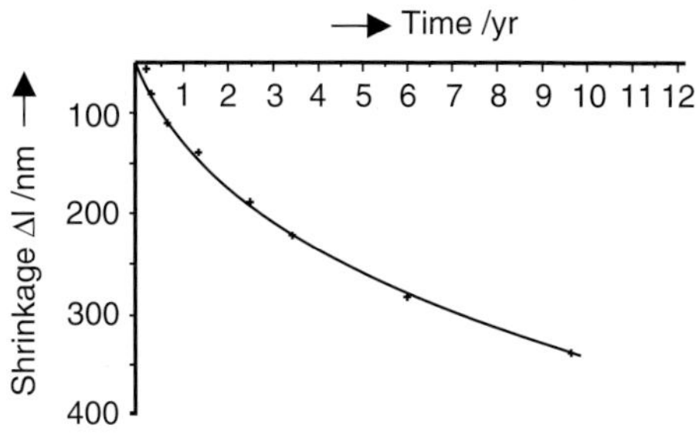

Fig. 4.64. Absolute longitudinal shrinkage of a Zerodur® bar with a length of 400 mm

where $\Delta l/l$ is the relative shrinkage during the time interval Δt. The observed shrinkage seems to run into a saturation area; in any case the ageing coefficient A reduces from year to year. The longitudinal ageing coefficient A depends on the annealing rates and the age of the Zerodur® sample, it is of the order of $A \leq 10^{-7}$ year^{-1}.

4.5.2 Radiation Stability

Due to their excellent physical properties, the low thermal expansion glass ceramics Zerodur® and Zerodur M® are approved materials for space applications like laser gyroscopes and mirror optics in telescopes [4.51]. These optical components are often directly exposed to space radiation environments defined by high temperatures, vacuum, and high-energy-particle and electromagnetic radiation of energies larger than 0.1 MeV. To study the change in evenness deduced by the hostile environment, the glass ceramics have been tested with simulated space radiation to evaluate their space usefulness. The radiation doses are calculated individually according to the experimental platform EURECA (European Retrievable Carrier) launched in 1992 [4.52].

Simulated Space Conditions

Radiation Environment. The type and intensity of the space environment consisting of particle and electromagnetic radiation depend on orbital height, inclination, and time. The magnetic field of the earth and the radiation field of the sun give rise to the so-called Van Allen belts, zones of high particle radiation. If possible, spacecraft avoid these regions. The Space Shuttle travels below the inner belt at 200–400 km. Like Shuttle flights, most of the space missions are scheduled for the so-called LEOs (low earth orbits) of a few hundred kilometers, where they are affected only by the rings of the inner Van Allen belt and of the outer atmosphere.

In LEOs the radiation environment involves the solar spectrum of photons, a few even more energetic photons from space, electrons, protons, and a few particles with higher mass, such as atomic oxygen from the outer atmosphere. NASA (National Aeronautics and Space Administration) and ESA (European Space Agency) use semi-empirical models to forecast accumulated doses for space missions. Doses must be calculated individually for each mission. Simulated irradiation doses were applied according to the experimental platform EURECA. The space samples were compared to ground samples after spending about one year on this carrier directly exposed to space at a height of about 500 km at low inclination. The calculations are based on the ESA model UNIRAD (program name) [4.53]. From the total doses listed in Table 4.6 it can be seen that most of the resulting defects are determined by the electrons. The dose of the protons makes only up for 10^{-3} of the electron dose. Therefore, only results on the irradiation of Zerodur® by electrons are reported here.

Table 4.6. Particle radiation environment: energies and doses for EURECA (calculated for one year)

Particle	Energy range in MeV	E_{average} in MeV	Dose in Gy Solmin-Solmax
e	$\leq$ 5	0.16	943–1722
p^+	$\leq$ 300	85	5–11
α	$\leq$ 40		
0°	5×10^{-6}		

Radiation Program. The high-quality properties of the low-expansion glass ceramics may be severely attacked by ionizing radiation. Electrons and protons of the above-mentioned energies interact almost exclusively with the electronic shells. An incoming primary particle causes a charge cascade along its path, bonds are broken, valences changed, and the glass network cannot perfectly relax back to its original status. However, primary particle energies are not high enough for the displacement of atoms or ions of the network or for considerable interactions with atomic nuclei.

The glass ceramics have densities around 2.5 g/cm^3, therefore, the penetration depth of space electrons and protons is only a few discoloring. Network modifications resulting in changed material properties affect the performance of glass ceramics only if they are exposed directly to space and mounted on the outside of the spacecraft. However, in this case as above only the first discoloring of the glass ceramics are concerned.

Electrons with adequate energies were provided by a Van de Graaf accelerator at the solid-state physics laboratories of Groningen University. According to the expected doses in Table 4.6, 1.3×10^3 Gy is the nominal dose for one year at mean sun activity. The realistic dose of 10^3 Gy produces only faint damages, if any. To investigate time dependence and the types of defects, 1, 5, 10, 10^3, and 10^4 years in space (10^3–10^7 Gy) were simulated.

The dose distribution of an energy spectrum was achieved by irradiating samples in two steps. The doses of the successive monoenergetic irradiations matched the integrated real particle flux and, on the other hand, the individual stopping power of Zerodur® (see Fig. 4.65). For example, the dose distribution for 1000 years in space was simulated by 20.5 Gy at 0.5 MeV and 1.3×10^4 Gy at 0.3 MeV.

Results

Intrinsic Defects. Radiation damage in low thermal expansion glass ceramics has been the subject of numerous investigations [4.54, 55]. The interaction of high-energy radiation with glassy materials is a rather complex phenomenon which depends mainly on the ionization radiation and the microstructure of the glass. One mode of forming paramagnetic defects, for example, involves

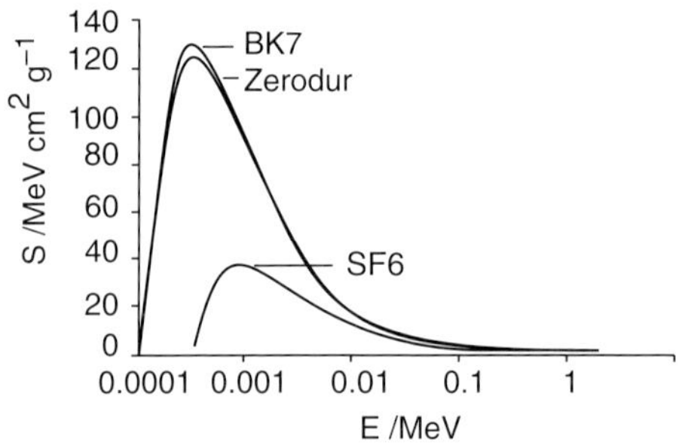

Fig. 4.65. Stopping power (S) of a borosilicate glass (BK 7), a lead silicate glass (SF 6), and Zerodur® for electrons

the interaction of ionizing radiation with the glass ceramic to form electrons and electron holes which are subsequently trapped at preexisting flaws in the glass network. The defect center most reported on in literature, i.e., the silicon E'-center, which is found in silicate glasses, especially in α-quartz, could not be observed [4.56]. These paramagnetic defects can be studied by electron paramagnetic resonance (EPR), an important tool for understanding defect generation.

Apart from other published experiments [4.57] related to EURECA, mission radiation and recording of EPR spectra have been done at room temperature (see spectrum 3 in Fig. 4.66). Spectra 1 and 2 result from samples irradiated at 77 K, recorded at 170 K and irradiated at 77 K, annealed at 440 K, and recorded at 170 K, respectively (published in [4.57]). The main, quite complicated signal around g_{free} consists basically of a variety of the so-called oxygen hole centers (OHCs) close to aluminum or silicon. aluminum has a natural abundance of 100% Al^{27} with I $= 5/2$, yielding a hyperfine coupling. In accordance with other publications it has been shown that all

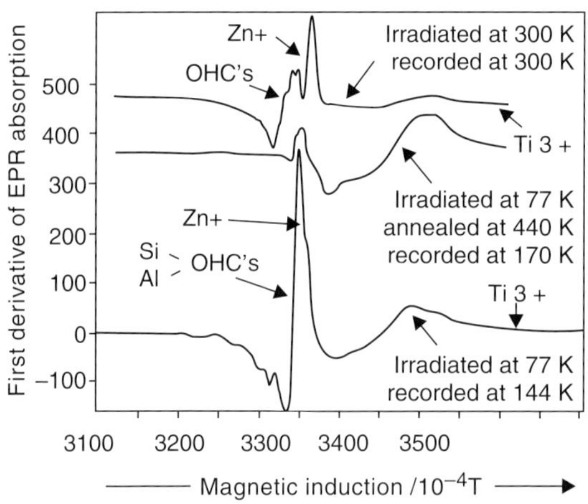

Fig. 4.66. EPR-signal intensity versus magnetic field strength – temperature dependence of the EPR-active centers in Zerodur®, created by electrons

the Al OHCs are situated in the h-quartz s.s. crystal phase of Zerodur®. The computer simulation of the spectrum needed no distribution in the spin Hamiltonians as is necessary in random networks.

Besides the hole centers the Zerodur® spectrum also contains electronic defects. These centers are designated as Zn^+ centers in the neighborhood of Al OHCs and Ti^{3+} centers at $g = 1.9$ (the g factor is well-known from the Zeeman effect), all acting as electron traps:

$$Zn^{2+} + e^- \rightarrow Zn^{1+} ,$$

$$Ti^{4+} + e^- \rightarrow Ti^{3+} .$$

They are paramagnetic centers.

As can be seen from Fig. 4.66 the Ti^{3+} center is very stable whereas the Zn^{2+} center disappeared after annealing at 440 K.

All these different signals are superimposed and differ in signal intensity with temperature during irradiation and recording.

Change of Physical Properties. According to the identified intrinsic defects, the physical properties of Zerodur® change macroscopically.

Optical darkening of Zerodur® α (cm^{-1}) is a linear function of an ionizing dose to doses significantly exceeding 10^3 Gy. Zerodur® colors more rapidly compared to glassy and crystalline materials and more severely than fused silica. These results are in accordance with previous findings [4.58]. The darkening of Zerodur® does not limit the material function as long as its optical properties are not used.

Density changes were measured with a density column. Changes could be detected only after high doses of 10^6 Gy electrons (0.8 and 1.35 MeV) had been applied. Zerodur® and Zerodur M® were compacted by $\Delta\rho \approx +0.013 \pm 0.005\,g/cm^3$ which is in accordance with the results published in [4.54] for 2 MeV electrons.

Optical figures were measured interferometrically before and after irradiation. Density changes of this order might be detrimental to optical mirror systems. Compaction in the top layer causes convex deformations of plane mirrors. Zerodur® and Zerodur M® samples with a diameter of 100 mm and a thickness of 5 mm were polished to $\lambda/50$ evenness at Carl Zeiss (Germany).

The results are shown in Table 4.7. The shape changes of Zerodur® and Zerodur M® are of the same order, but distinctly smaller in Zerodur M®.

Table 4.7. Change in the evenness of Zerodur® and Zerodur M® EURECA (diameter 100 mm, thickness 5 mm)

Electron dose Gy 0.8 MeV	+0.3 MeV	Equivalent to time in orbit	Zerodur®	Zerodur M®
20.5	$+1.3 \times 10^4$	10 years	$-$ 1.6 μm	0.8 μm
4.1×10^3	$+1.3 \times 10^6$	10^3 years	-12.5 μm	10.3 μm

X-ray diffraction studies revealed there were no changes exceeding 1% in the crystal phase content.

Conclusions

The results mentioned above show that the hostile radiation fields in the inner and outer region of the Van Allen belt in space can be simulated. The radiation doses are no limitations to any space applications of Zerodur®.

Only very high doses increased by a factor of 1000 over a real one-year dose in the EURECA mission show slight changes in shape performance, the main physical property claimed for mirror blanks in space telescopes. If a material is needed which is able to comply with new tasks in a more extreme radiation environment, Zerodur M® offers more advantages.

The final realistic test results on EURECA have been available since 1995. They include the non-simulated space conditions such as isotropic incidence of radiation, the real dose rates, the special temperature of the mission, and the synergetic effects of the different radiation species.

4.5.3 Chemical Treatment

Normally a chemical treatment of Zerodur® is carried out for one of the following purposes:

- in order to get rid of surface notch tensions after production by removing a surface layer (mostly in the perimeter area of a Zerodur® disk) with an etching procedure,
- in order to remove optical mirror coatings from Zerodur® mirror substrates before re-coating.

In the first case a chemical attack of the Zerodur® material is desired. In the second case the chemical treatment should not attack the Zerodur® material in a noticeable way, only the coating (normally an aluminum layer) is to be dissolved.

In the following the two cases are described in more detail.

Reduction/Removal of Surface Notch Tensions by Etching

Hydrofluoric-acid/hydrochloric-acid etching baths are suitable for removing noticeable amounts of material from Zerodur® in warrantable times. Microcracks are healed up or removed by this process. Due to the use of hydrochloric acid in the etching bath the formation of fluoride layers which are difficult to dissolve on the surface is not possible; these fluoride layers would affect the etching process.

The stock removal per surface and unit of time at a given temperature changes depending on the composition of the etching bath. The treatment

Table 4.8. Treatment periods for a stock removal of 0.05 mm for three different etching baths

Etching bath composition		Mean time for a stock removal of 0.05 mm per surface
2.0 volume parts HF 1.0 volume parts HCl	40% 32%	3.0 min
2.0 volume parts HF 1.0 volume parts HCl 1.5 volume parts H_2O	40% 32% dist.	7.5 min
2.0 volume parts HF 1.0 volume parts HCl 3.0 volume parts H_2O	40% 32% dist.	17.5 min

Etching bath temperature: 20 °C

period for a stock removal of 0.05 mm per surface at 20 °C is given in Table 4.8 for three different etching baths. The function of time of the stock removal is linear in the case of the various etching baths.

The surfaces have to be cleaned (grease-free) prior to etching. The typical micro-roughness is in the range of 3–4 μm. The surfaces have a semitransparent appearance after etching.

The description of the etching method is only meant to serve as an indication. In all individual cases, the etching time required has to be determined as a function of the initial state of the surface to be treated.

Methods for Separating Mirror Coatings from Zerodur® and for Cleaning

The reflection power of optical mirror coatings on mirror supports may decrease with time. In these cases the mirror coating must be removed, and the mirror support must be re-coated or vapor-deposited. The coatings normally consist of aluminum. For the separating and cleaning of these coatings on Zerodur®, two methods have been developed which are to be applied at room temperature.

With both methods the Al coatings are removed so carefully that these processes may be repeated at least a hundred times. Changes in the surface (micro-roughness or the like) due to chemical reactions, which may adversely affect the use of the mirror support after re-coating, do not occur. Separation is supported by a gentle mechanical movement such as smooth wiping with a soft cloth in order to avoid surface injuries and scratches.

Method 1:

- exposure to aqueous sodium hydroxide solution of 8% containing 0.3% of an alkaline wetting agent, exposure time 10 min;

- rinsing with water, three times;
- exposure to hydrochloric acid of 10%, exposure time 10 min;
- rinsing with water, twice;
- rinsing with distilled or exchanged water, twice;
- drying.

Method 2:

- exposure to hydrochloric acid of 10% containing 0.5% of copper sulfate and 0.3% of an acid wetting agent, exposure time 30 min;
- rinsing with water, three times;
- exposure to caustic soda solution of 10%, exposure time 10 min;
- rinsing with water, twice;
- exposure to hydrochloric acid of 10%, exposure time 5 min;
- rinsing with water, twice;
- rinsing with distilled or exchanged water, twice;
- drying.

4.6 Applications of the Glass Ceramic Zerodur®

Alfred Jacobsen, Thomas Marx

The specific properties of Zerodur® have been described in Sect. 4.1. They are extraordinary by themselves, and, especially combined with the material's ability to allow production in large volumes and its machinability, they have resulted in new applications and, in particular, in great progress and the achievement of new possibilities in certain technologies.

Most applications take advantage of the negligibly small coefficient of thermal expansion and its homogeneity over the entire volume. This property provides for stability in shape and volume if the piece is exposed to temperature changes and temperature gradients. This behavior is a requirement, in particular, for mirror substrates in precision reflective optics supports, frames, or scales and gauges.

4.6.1 Reflective Optics

Imaging optical systems with a very large diameter are built with reflective optics, because refractive optics pose problems of high light absorption with increasing glass thickness. The reflection of the light occurs on a thin metal layer that has to be applied to a smooth substrate. Zerodur® is the material of choice for mirror substrates due to its above-described properties since it can be precision machined and polished in combination with the advantage of zero thermal expansion which eliminates the influence of temperature fluctuations on the quality of the image during the observations.

Astronomical Telescopes

The development of astronomical telescopes started around 1608 when Galileo Galilei discovered the four largest Jupiter moons under 30 times magnification. Like Johannes Kepler, he used refracting telescopes. Isaac Newton developed the first reflecting telescope with the large primary mirror reflecting and focusing light beams perpendicular to the side with a secondary mirror.

The mirror optics for large telescopes, which were developed later on by Cassegrain, utilized a large concave primary mirror which collects as much light as possible, and a convex secondary mirror. The light beams are focused with the secondary mirror through the vapor hole of the primary mirror in the "shadow" of the secondary with the focusing spot moving with the telescope. The possibility of a focus in a fixed location can be realized by lateral diversion of the beam with a plano mirror allowing the installation of large detectors and sensors.

Contrary to common belief, large magnifications only play a minor role in the observation of cosmic objects. A large diameter of the optic is more important to collect more light beams and also to improve the resolution of the image. The size of the mirrors, however, poses a significant challenge to their stiffness and the cost-efficient resolution of weight problems for the optical elements in the telescope that have to track the "movement of the star" during the observation with utmost precision.

During their 25-year history, mirror substrates manufactured from Zerodur® have pursued all developmental directions for large telescopes. The first milestone was a stiff mirror blank made from Zerodur® with a diameter of 3.6 m and a weight of 14.5 t for the Max Planck Institute of Astrophysics in Heidelberg, Germany, which erected a telescope on the Calar Alto in the Sierra Nevada in southern Spain. Its average thickness is approximately 595 mm with the mirror surface having a radius of curvature of approximately 24.5 m. At the time, the precision-machined and polished surface marked a performance record since the resolution in the range of arc seconds surpassed that of existing larger telescopes (Palomar Telescope/USA, of diameter 5 m; Zelentchukskaya Telescope/Russia, of diameter 6 m).

The European Southern Observatory (ESO) has started operations of a telescope on La Silla in Chile that has a meniscus-shaped monolithic Zerodur® mirror blank of 3.6 m in diameter, a thickness of only 240 mm, and a weight of 6.5 t (see Fig. 4.67). The reduced weight has significant advantages for a lighter and more cost-efficient design of the telescope.

On the other hand, this substrate is not stiff enough to prevent deformation under its own weight. A new technology was utilized and tested which gave the telescope its name. The system of active optics developed by Dr. R. Wilson was tested with the "New Technology Telescope" (NTT). It facilitates a continued controlled adaptation of the mirror surface to the ideal shape by selectively diverting the light from a reference star to the side of the telescope. This allows for computer-aided activation of 78 actuators

Fig. 4.67. ESO's New Technology Telescope (3.5-m Zerodur® mirror) obtains the highest performance by using active and adaptive optics

supporting the mirror in a way that makes possible a maximum image quality. This system offers the possibility to correct "deformations" caused by the mirror's own weight and potentially existing permanent residual errors of the polished surface contour, but it also corrects wavefront errors resulting from the "seeing" of the telescope due to air convection.

Another possibility for optimizing the diameter-to-thickness ratio of a substrate for a mirror of large diameter is being realized by the California Association for Research in Astronomy (CARA) with the Keck Telescope on Mauna Kea in Hawaii. A total of 36 hexagonal Zerodur® mirror segments with a thickness of 75 mm form the primary mirror of 10 m diameter (see Fig. 4.68).

Fig. 4.68. One of 36 blanks made of Zerodur® for the 10-m Keck Telescope

A sophisticated computer-controlled support system keeps the mirror segments in the desired position. The finishing of the off-axis aspheric segments is performed by stressed mirror polishing, a newly developed process for which the initially round meniscus-shaped blanks of 1.9 m diameter are mechanically deformed in such a way that a desired aspheric surface results after the stress is relieved.

Even prior to the completion of the first Keck Telescope, a second one has been erected to double the amount of light collected when the output of both telescopes is brought to interference. At the same time, the distance and location between the telescopes will improve the resolution of the total system.

A new record was achieved by the Very Large Telescope (VLT) of the ESO. Four telescopes with mirror diameters of 8.2 m each have been installed on the Cerro Paranal in Chile. Together they represent a useful mirror surface equating to that of a mirror of diameter 16 m.

The required meniscus-shaped monolithic Zerodur® mirror substrates weighing 23 t each with 177 mm thickness represented a big challenge for glass melting, thermal treatment, handling, and transport (see Fig. 4.69, refer also to Sect. 4.3.1).

The glass ceramic Zerodur® is used for primary and secondary mirrors in a variety of other astronomical telescopes. All these mirror substrates are finished with aspheric surface contours. The qualities achieved range around a 1/50 wave deviation from the target surface figure or approximately 10 nm.

Fig. 4.69. One of four monolithic mirrors of diameter 8.2 m for the Very Large Telescope of the ESO

Only this precision allows the record breaking achievements of modern astronomy.

Mirror Optics in Space

Zerodur® mirror substrates are also used in space applications for many other optical systems.

The requirements for the precision of the optical image are so high that temperature fluctuations from the changing solar irradiation must not have an influence on the mirror surfaces. Materials with extremely small coefficients of thermal expansion are as much a requirement as the precision polishability.

A radiometer is installed in the European weather satellite METEOSAT at a geostationary orbit of 36 000 km altitude over equatorial Africa imaging the Earth's surface with a resolution of 2.5 km in order to monitor weather conditions. The camera system with an aperture of 400 mm has to resist very large temperature fluctuations during the change from direct solar irradiation to the shadow of the Earth and must also operate for many years without maintenance.

The French Earth reconnaissance satellite (SPOT) has been performing a similar task orbiting our planet since 1986 and delivering snapshots of utmost clarity of the Earth's surface with a resolution of 10–20 m. The Zerodur® mirror blank for the primary mirror was milled from the back side to reduce the weight by 60% (see Fig. 4.70). Higher lightweighting factors for space

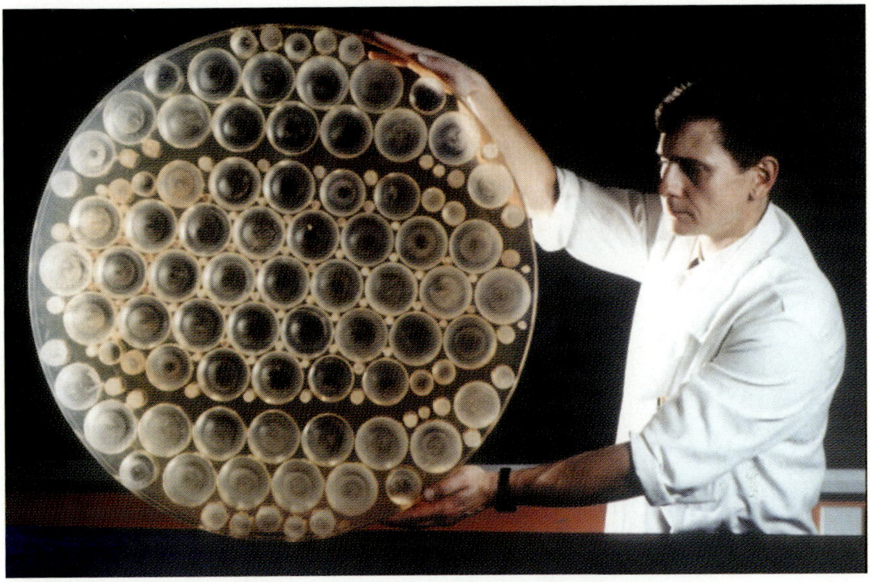

Fig. 4.70. A lightweight mirror blank made by the R.E.O.S.C. Company in France, for the Earth observation satellite SPOT

optics or secondary telescope mirrors can be achieved by fabricating fused structures (refer to Sect. 4.3.2).

Another Zerodur® mirror optic is contained in the comet space probe GIOTTO, which photographed and analyzed Halley's Comet in 1986 while passing by at a distance of 14 500 km from the comet's vapor. The space probe has subsequently registered various cosmic events on its way through the solar planetary system. An encounter with the comet Grigg–Skjellerup in the year 1992 even allowed an observation for several weeks.

X-Ray Telescopes

The astronomical investigations of space are not restricted solely to optical radiations. Space technology, starting with ballistic rockets and, in particular, satellite technology, have enabled astronomers to study X-ray sources which are the result of interstellar catastrophical events such as the collapse of a solar system. These observations are only possible outside the atmosphere of the Earth (see Sect. 4.3.3).

The German X-ray satellite telescope ROSAT was placed in orbit in 1990. It is equipped with a four-mirror system of parabolic and hyperbolic Zerodur® cylinders nested into one another (see Fig. 4.71). The inner surfaces of these reflectors were ground and polished by Carl Zeiss, Germany. The residual surface roughness is less than 0.2 nm RMS. An angular resolution of 3.3 arc

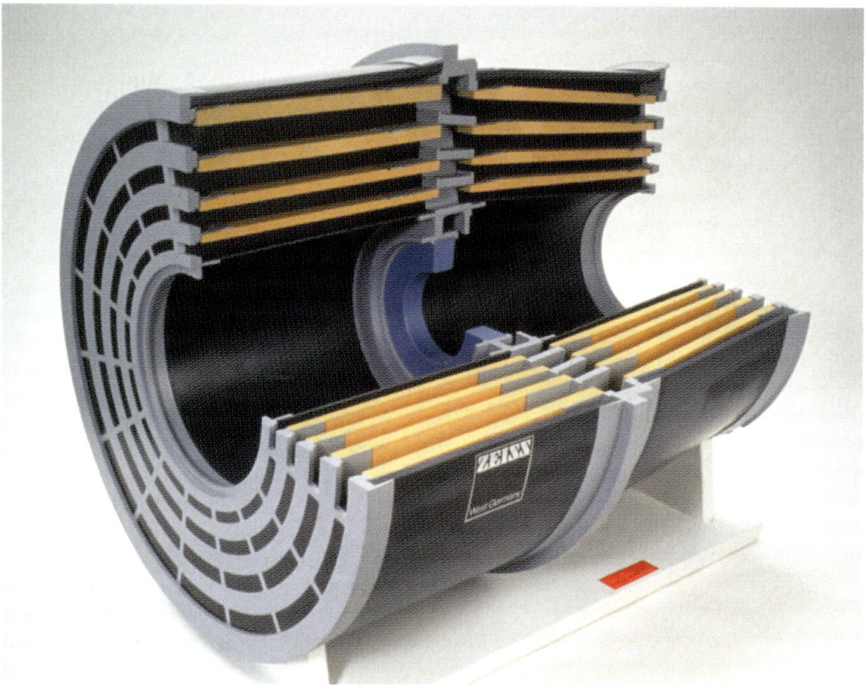

Fig. 4.71. Four parabolic and hyperbolic Zerodur®cylinders are nested into one another for the German spaceborne X-ray telescope ROSAT

seconds was achieved for the telescope due to the precise fabrication and mounting of the mirrors.

The task of this X-ray satellite was initially to scan space systematically for X-ray sources followed by a pointed spectroscopic analysis as well as a study of changes in the characteristics of single X-ray sources in space over time.

The US Space Agency, NASA, is also building a new X-ray satellite telescope, Advanced X-Ray Astrophysics Facilities (AXAF), for the spectral region of 0.1–10 KeV for an Earth orbit at 600 km altitude (Fig. 4.72). The launch is planned for 1998. This Wolter telescope, with six paraboloids and hyperboloids each nested into one another, consists of Zerodur® cylinders with diameters between 600 and 1274 mm, whose fabrication required the development of many new techniques (refer to Sect. 4.3.3).

Mirror Substrates in Microlithography

Mirror systems with the glass ceramic Zerodur® can produce resolution capabilities of 500 line pairs per mm for microlithographic processes. This impressive performance, which allows a resolution of line widths of approximately

Fig. 4.72. One of twelve thin-walled conical Zerodur® cylinders of diameters up to 1274 mm for the AXAF X-ray telescope at NASA

1 μm, is required for the technology of microchip production. It is used to project the desired semiconductor structures onto the silicon wafer. The Silicon Valley Group, formerly Perkin Elmer Corp. USA, has significantly influenced the race for powerful microchips, for example, with its Micralign® series, by using Zerodur® mirror substrates (see Fig. 4.73). This system allows not only a high resolution and accuracy of superposition but also a high throughput. Pieces of equipment produce 120 wafers per hour at a size of 100 mm and 100 wafers of 150 mm. Similar achievements have also been accomplished by the Japanese companies Canon and Nikon.

Mirror Substrates for Optical Metrology

Interferometers are optical devices for length measurements in the range of the wavelengths of visible light to measure refractive indices, homogeneity, and strain in transparent materials as well as the surface figure and transmitted wavefront in finished optics. For instance, light transmitting through a glass piece to be measured is brought to interference with a corresponding reference beam in a Fizeau interferometer (see Fig. 4.74).

Zerodur® is used as a mirror blank for the optics and represents, to a large degree, the quality-determining factor in the overall device.

Fig. 4.73. Precise contour measurements of a mirror blank for microlithographic processes for powerful microchips

Fig. 4.74. Zerodur® mirrors form the reference surface in a Fizeau interferometer of highest accuracy

4.6.2 Laser Gyroscopes

A very large area of application for the glass ceramic Zerodur® is the technology for modern navigation of space rockets, aircraft, and ships or the positioning of telescopes, antennas, or satellites. The negligibly small coefficient of thermal expansion and the extremely small helium permeability of the material are of particular importance for this application.

Triangular or square Zerodur® blocks with a ring-shaped cavity and attached reflector are used as the thermally stable frame for the He-Ne laser beams propagating in opposite directions (see Fig. 4.75). The two laser beams are at the same frequency and in resonance if the ring-shaped laser gyroscope is in a stationary position. If the gyroscope is rotated around its axis perpendicular to the plane of the ring, one beam has a slightly longer, the other a slightly shorter, length of travel (Sagnac effect). Consequently, the resonance frequencies of both beams change – one becomes larger, the other becomes smaller, resulting in a resonance pattern. The sensor shows a value that is proportional to the angular velocity of the gyroscope.

For navigation technology the conventional electro-mechanical gyroscope is to be replaced by a system with digital analysis without mechanically moving parts.

The advantage of Zerodur® lies in the above-mentioned low permeability of the material for helium atoms which is of particular importance for

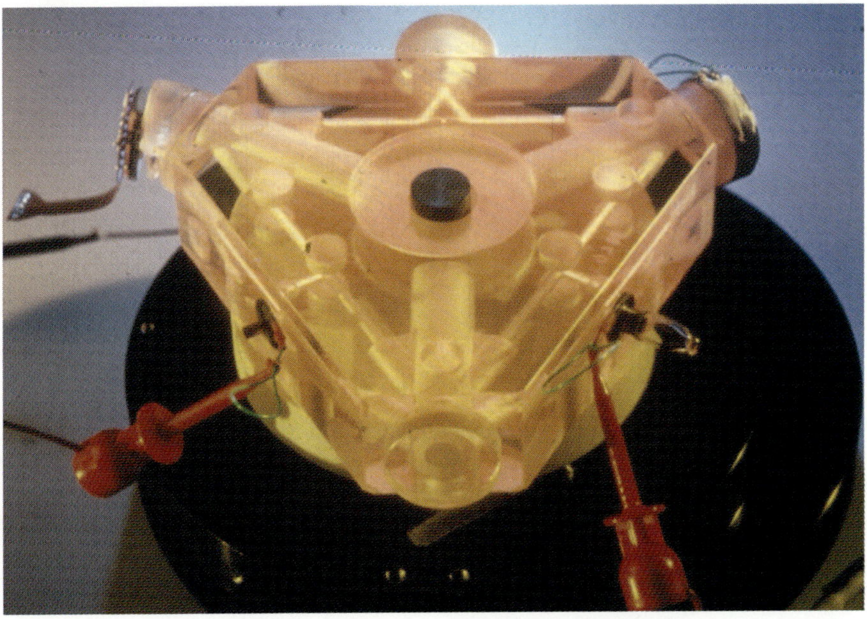

Fig. 4.75. Low helium permeability makes Zerodur® the preferred material for laser gyroscopes in modern navigation technology

the long-term stability. The remarkable properties of the sensor are a measurement range from 10^{-6}–$10^{30}\,s^{-1}$, an angular resolution of approximately 2 s, a zero point stability of 10^{-3} °/h, and a scale factor stability of approximately 3 ppm. Therefore, the navigation system has been changed to this new method for the European Airbus, Boeing aircrafts, for submarines, for the space rocket ARIANE, and for the stabilization of satellites.

G Ring Laser Gyroscope

In 2001, the so-called G ring laser gyroscope was taken into service at the Wettzell Fundamental Research Station with the purpose of precisely measuring the Earth's rotation.

The frequency difference occurring with the normal variations in the Earth's rotation in the course of a day is $10^{-6}\,s^{-1}$ out of a frequency of $300\,s^{-1}$. To be able to record such minute differences, the sensitivity of the G ring laser gyroscope has to be more than a thousand times higher than that of "normal" gyroscopes used, for example, in navigation systems.

The optical path difference as a measure of the rotation speed of the reference system is proportional to the surface enclosed in the ring laser, i.e., the bigger the circuit traveled by the rays, the higher the accuracy of the measurements (see Fig. 4.76). Therefore, the size of the ring laser is a crucial factor. The second prerequisite is the absolute stability of the measurement array. Any change would influence the travel time of the laser light and thus falsify the measurement results.

For these reasons, the G ring laser was equipped with a giant Zerodur® disk, 4.25 m in diameter, 25 cm thick and weighing 10 t (Fig. 4.77). The already very low thermal expansion of the material Zerodur® was further re-

Fig. 4.76. Sketch of the G ring laser gyroscope

Fig. 4.77. Zerodur disk for the Wettzell Fundamental Research Station

duced by post-ceramization, resulting in a thermal expansion coefficient of 60 nm/K. With this extremely stable base and 16 square meters of surface traveled, the Wettzell ring laser is the largest and most accurate gyroscope in the world.

4.6.3 Precision Engineering

There is another very broad application area for the glass ceramic Zerodur® in precision engineering.

Since most materials (metals, glasses, crystals, plastics, etc.) expand and contract with temperature variations, thus changing their shape, substantial precautions must be taken to keep the temperature constant (air-conditioned and temperature-stabilized rooms, etc.), or long waiting periods have to be accepted to achieve a temperature stabilization.

The glass ceramic Zerodur® does not change its shape due to the negligibly small coefficient of thermal expansion, even if the ambient temperature changes between 0 and 100 °C. This is why, for instance, the resonators of CO_2 lasers in high-power laser technology are being kept in that exact position, even if the ambient temperature or the temperature in parts of the lasers changes, by using multiple piece rods of Zerodur® of up to 20 m in length.

A particular technical progress has been achieved by the company Wegmann-Baasel, Germany, which has significantly improved the long-term stability of the focal point for the CO_2-laser beam by using temperature-stable spacer plates in the resonator (see Fig. 4.78). This results in more reliable cutting and welding processes combined with economic advantages offered by high feed rates.

Massive vibration-free tables of Zerodur® or plates of Zerodur® on metal tables are the temperature-stable bases for the high-precision micropositioning of optical components in laboratories where the temperature cannot be kept constant at all or only with great difficulty (see Fig. 4.79).

Scales and gauges from the glass ceramic Zerodur® are also in widespread use, since they are the temperature-independent standard for length measurement. This usage takes advantage of another property of Zerodur® – the length stability over time.

The micropositioning of microchips for the manufacturer of wafers is a typical one among the many applications (see Fig. 4.80). A substrate part made of Zerodur® as the stage can position the wafer in six directions with extreme precision since certain zones of its surface are equipped with precision polished areas and mirrors for sensors. This application is typical as the homogeneous material can be machined to extreme precision; it does

Fig. 4.78. In high-energy lasers, Zerodur® distance spacers keep the optical elements in exact position to improve the long-term stability of the focal point of the laser beam

Fig. 4.79. A temperature-stable table for the high-precision micropositioning of optical components

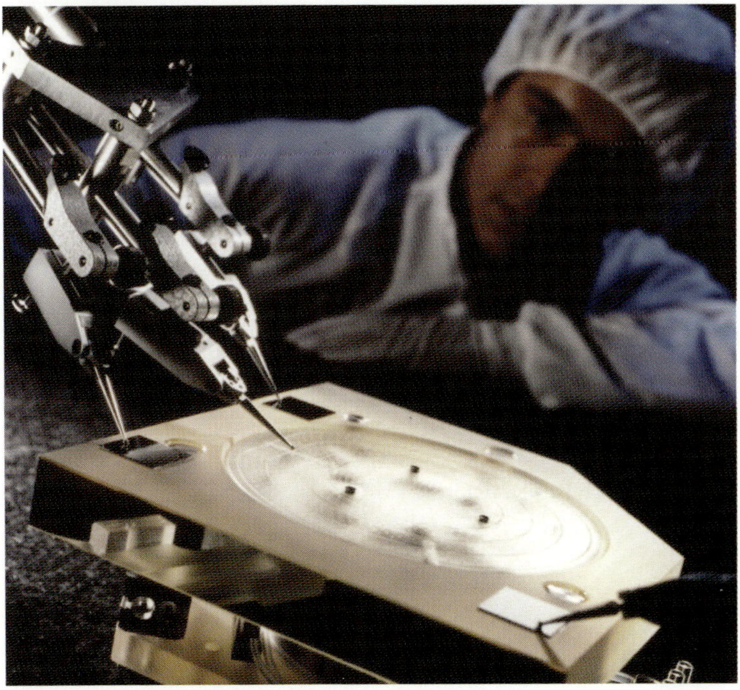

Fig. 4.80. Supports made of Zerodur® for highest precision positioning of wafers in microlithography at Philips, NL

not change with temperature fluctuations and nevertheless meets the high requirements of mass production in a rough manufacturing environment.

Again, using the temperature-stability characteristic, Zerodur® is increasingly being designed into systems such as analytical instruments for medical and other purposes or as holding fixtures for elements that require stable positioning such as sensors.

The low thermal expansion also makes Zerodur® a material with a very high thermal shock resistance. This feature in combination with its transparency offers interesting design options. They are currently being exploited, for instance, in the packaging industry where plastic bags can be sealed with a Zerodur® bar while the quality of the seal is inspected through the bar. Similar possibilities may exist in speciality plastic injection moldings and other industries.

General Reading

D. Fischer: "Auf dem Wege zu neuen Großteleskopen", Sterne und Weltraum **11/88**, 654–657, 27. Jahrg.

A. Jacobsen: "High flying Zerodur®", Schott Information **3**, 2–9 (Mainz 1987)

A. Jacobsen, R. Müller: "50 Quadratmeter höchste Präzision", Sterne und Weltraum **1/90**, 20–25, 29. Jahrg.

O. Lindig, W. Pannhorst: "Thermal expansion and length stability of Zerodur® in dependence on temperature and time", Appl. Opt. **24**(20), 3330–3334 (1985)

T.A. Marx: "Advanced Processing of Zerodur®", Glass Ceramic, Proc. SPIE **1535** (1991)

F. Merkle, M. Schneermann: "Das ESO Very Large Telescope", Sterne und Weltraum **9/86**, 460–466, 25. Jahrg.

R. Müller, H. Hoeness, T. Marx: "Spincast Zerodur mirror substrates of the 8 meter class and lightweighted substrates for secondary mirrors", Conference on Astronomical Telescopes and Instrumentation for the 21st Century, Proc. SPIE **1236** (1990)

J. Nelson: "The Keck Telescope", American Scientist **77**, 170–175 (1989)

E. Schuster: "Astronomie in neuen Dimensionen", Schott Information **48**, 4–9 (Mainz 1988)

Schott Glaswerke: Schott Information **2** (Mainz 1986)

References

4.1 W. Sack, H. Scheidler: "Einfluß der Keimbildner TiO_2 und ZrO_2 auf die sich ausscheidenden Kristallphasen bei der Bildung von Glaskeramik", Glastechn. Ber. **39**, 126–130 (1966)

4.2 M. Tashiro, K. Takagi: "Method of preventing scum formation in glass melts, and glass-ceramic products", US Patent 3 282 712 (Nov. 1, 1966)

4.3 J. Petzoldt: "Metastabile Mischkristalle mit Quarzstruktur mit Oxidsystem Li_2O-MgO-ZnO-Al_2O_3-SiO_2", Glastechn. Ber. **40**, 385–396 (1967)
4.4 H. Saalfeld: "Struktur und Ausdehnungsverhalten von Li-Al-Silikaten", Ber. Dt. Keram. Ges. **38**, 281–286 (1961)
4.5 W. Schreyer, J.F. Schairer: "Metastable solid solutions with quartz-type structures on the join SiO_2-$MgAl_2O_4$", Z. Kristallogr. **116**, 60–82 (1961)
4.6 O.W. Flörke: "Kristallisation und Polymorphie von $AlPO_4$ und $AlPO_4$-SiO_2-Mischkristallen", Z. Kristallogr. **125**, 134–146 (1967)
4.7 J. Petzoldt: "Der Einbau von P_2O_5 in metastabile Mischkristalle mit Quarzstruktur des Grundsystems Li_2O-MgO-ZnO-Al_2O_3-SiO_2", Glastechn. Ber. **41**, 181–189 (1968)
4.8 R.W. Dietz, J.M. Bennett: "Smoothness and thermal stability of Cer-Vit® optical material", Appl. Opt. **6**, 1275–1276 (1967)
4.9 R.C. Monnier: "Fabrication of a 104 cm mirror from Cer-Vit® low expansion material", Appl. Opt. **6**, 1437–1440 (1967)
4.10 D.A. Duke, G.A. Chase: "Glass-ceramics for high precision reflective-optic applications", Appl. Opt. **7**, 813–817 (1968)
4.11 J. Petzoldt: "Transparente Glaskeramiken mit einem thermischen Ausdehnungskoeffizienten von $0 + 1,5$ mal 10^{-7}/Grad C, der im Bereich von -30 bis $+70\,^\circ$C wenig temperaturabhängig ist", BD Patent 1902432 (June 30, 1977)
4.12 J. Petzoldt: "Untersuchungen an transparenten, metastabile Quarz-Mischkristalle enthaltenden Glaskeramiken", Glastechn. Ber. **43**, 127–137 (1970)
4.13 Schott Glaswerke Mainz: "Zerodur® – precision from glass ceramics", Brochure No. 10041 (1991)
4.14 Schott Glaswerke Mainz: "Glass ceramic Zerodur M®", Product Information No. 10008 (Schott Glas, Mainz 1989) available on request
4.15 H. Richter, G. Kleer: "Untersuchungen zur Charakterisierung des Festigkeitsverhaltens von Zerodur®", Fraunhofer-Institut für Werkstoffmechanik, Internal Report (1983)
4.16 D.S. Tucker: "Dynamic fatigue of a lithia-alumina-silica glass-ceramic", J. Am. Ceram. Soc. **73**, 2528–2530 (1990)
4.17 Schott Glaswerke Mainz: "Design strength of optical glasses and Zerodur® ", Technical Brochure (1991)
4.18 G. Exner, O. Lindig: "Bestimmung des Widerstandes der Spannungsrißkorrosion n an Borosilicatglas Duran®", Glastechn. Ber. **55**, 107–117 (1982)
4.19 F. Ayer: "Materials for space optics generic requirements", Presentation at ROFT-Program meeting, Daytona Beach (Jan. 31, 1984)
4.20 Schott Glaswerke Mainz: "Schott optical glass", Catalogue No. 10000e (1992)
4.21 U. v. Alpen, E. Schönherr, H. Schulz, G.H. Talat: "β-eucryptite – a one-dimensional Li-ionic conductor", Electrochim. Acta **22**, 805–807 (1977)
4.22 J.E. Shelby: Measurements by Shelby; private communication to O. Lindig
4.23 J. Petzoldt: Internal Report (1972)
4.24 S.J. Bennett: "An absolute interferometric dilatometer", J. Phys. **E10**, 525–530 (1977)
4.25 W. Gorski: "Längenänderung von Zerodur®", Jahresber. Phys. Techn. Bundesanstalt, Braunschweig, 159–160 (1978)
4.26 O. Lindig: private communication (1980)
4.27 O. Lindig, W. Pannhorst: "Thermal expansion and length stability of Zerodur® in dependence on temperature and time", Appl. Opt. **24**, 3330–3334 (1985)
4.28 J.J. Schaffer, H.E. Bennett: "Effect of thermal cycling on dimensional stability of Zerodur® and ULE", Appl. Opt. **23**, 2852–2853 (1984)

4.29 S.F. Jacobs, S.C. Johnston, G.A. Hansen: "Expansion hysteresis upon thermal cycling of Zerodur®", Appl. Opt. **23**, 3014–3016 (1984)
4.30 O. Lindig, W. Pannhorst: "Length changes in Zerodur® caused by thermal cycling", Paper presented at 87th Annual Meeting of Am. Ceram. Soc., Cincinnatti (1985)
4.31 O. Lindig, W. Pannhorst: "Glass-ceramic with specific thermal expansion behavior", US Patent 4,851,372 (July 25, 1989)
4.32 S.F. Jacobs, S.C. Johnston, J.M. Sasian, M. Watson, J.D. Targove, D. Bass: "Surface figure changes due to thermal cycling hysteresis", Appl. Opt. **26**, 4438–4442 (1987)
4.33 S.F. Jacobs, D. Bass: "Improved dimensional stability of Corning 9600 and Schott Zerodur® glass ceramics", Appl. Opt. **28**, 4045–4047 (1989)
4.34 R. Haug, A. Klaas, W. Pannhorst, E. Rodek: "Length variation in Zerodur M® in the temperature range from −60 °C to +100 °C", Appl. Opt. **28**, 4052–4054 (1989)
4.35 K.P. Birch, P.T. Wilton: "Thermal expansion data for Zerodur® from 247 to 373 K", Appl. Opt. **27**, 2813–2815 (1988)
4.36 H. Bach, N. Neuroth (Eds.): *The Properties of Optical Glass*, Schott Series on Glass and Glass Ceramics – Science, Technology, and Applications (Springer, Berlin, Heidelberg 1995)
4.37 R.W. James: *The Optical Principles of the Diffraction of X-Rays* (Bell, London 1967)
4.38 H. Wolter: "Spiegelsysteme streifenden Einfalls als abbildende Optiken für Röntgenstrahlen", Ann. Phys. **10**, 94 ff. (1952)
4.39 K. Beckstette, B. Aschenbach, M. Schmidt: "Assembly of the ROSAT flight X-ray mirrorsystem", SPIE Proc. **982/01**, 2–9 (1988)
4.40 J. Trümper: "Der Röntgensatellit ROSAT", Phys. Bl. **46** (1990)
4.41 J. Trümper: "ROSAT first light", Inst. Doc. (1990)
4.42 R. Wilson: "'First light' in the NTT", ESO Messenger **56**, 1–5 (1989)
4.43 E.-D. Knohl: "Fabrication of large and fast mirrors with extraordinarily smooth surfaces", in *The Next Generation Space Telescope Workshop* (Space Telescope Science Institute, Baltimore 1989) pp. 209–216
4.44 S.J. Bennet: "An absolute interferometric dilatometer", J. Phys. E **10**, 525–530 (1977)
4.45 W. Gorski: "Längenänderung von Zerodur", in *Jahresbericht der Physikalisch-Technischen Bundesanstalt* (Braunschweig 1978) pp. 159–160
4.46 S.F. Jacobs, S.C. Johnston, G.A. Hansen: "Expansion hysteresis upon thermal cycling of Zerodur", Appl. Opt. **23**, 3014–3016 (1984)
4.47 O. Lindig, W. Pannhorst: "Thermal expansion and length stability of Zerodur in dependence on temperature and time", Appl. Opt. **24**, 3330–3334 (1985)
4.48 W. Pannhorst, R. Haug, E. Rodek, K. Stetter: "Hysteresis effects in low expansion glass-ceramics", J. Non-Cryst. Solids **131–133**, 488–491 (1991)
4.49 K.P. Birch, P.T. Wilton: "Thermal expansion data for Zerodur from 247 K to 373 K", Appl. Opt. **27**, 2813–2815 (1988)
4.50 F. Bayer-Helms, H. Darnedde, G. Exner: "Längenstabilität bei Raumtemperatur von Proben der Glaskeramik Zerodur", Metrologia **21**, 49–57 (1985)
4.51 J. Bourrieau, M. Romero: "Effect of space charged particle environment on optical components and materials", ESA Spacecraft Mater. in Space Environment, 275–285 (1979)
4.52 E. Rädlein, G.H. Frischat, B. Speit: "Simulation of the interaction of space radiation with glasses and glass ceramics", in *The Physics of Non-Crystalline Solids* (Taylor & Francis, London 1992) pp. 261–265

4.53 E.J. Daly: "The UNIRAD system for radiation prediction ESTEC TN" (1986) and E.J. Daly et al.: "ESABASE/RADIATION update note", ESA/ESTEC/WMA (1987)
4.54 P.L. Higby, E.J. Friebele, C.M. Shaw, M. Rajaram: "Radiation effects of the physical properties of low expansion-coefficient glasses and glass ceramics", J. Am. Ceram. Soc. **71**, 796–802 (1988)
4.55 E. Rädlein: "Simulation der Wechselwirkung von Weltraumstrahlung mit Glas und Glaskeramik", PhD Thesis (Technical University Clausthal 1991)
4.56 E.J. Friebele, D.L. Griscom: "Radiation Effects in Glass", in *Treatise on Materials Science and Technology*, Vol. 17, Glass II (Academic Press, New York 1979) pp. 257–351
4.57 T.E. Tsai, E.J. Friebele, D.L. Griscom, W. Pannhorst: "Radiation induced defect centers in glass ceramics", J. Appl. Phys. **65**, 507–514 (1989)
4.58 B.C. Passenheim, C.E. Mallon: "Nuclear radiation vulnerability of ring laser gyroscope", SPIE **157**, 205–211 (1978)

List of Contributors

Manfred-Josef Borens
Schott AG, Mainz [1]
E-mail: manfred-josef.borens@schott.com

Roland Dudek
Schott AG, Mainz [1]
E-mail: roland.dudek@schott.com

Torsten Gabelmann
Schott AG, Mainz [1]
E-mail: torsten.gabelmann@schott.com

Helga Götz
Schott AG, Mainz [1]
E-mail: helga.goetz@schott.com

Ruban Harikantha
Schott AG, Mainz [1]
E-mail: ruban.harikantha@schott.com

Reiner Haug
Schott AG, Mainz [1]
retired

Wilfried Heimerl
Schott AG, Mainz [1]
retired

Rüdiger Hentschel
Schott AG, Mainz [1]
E-mail: ruediger.hentschel@schott.com

Hartmut Höneß
Schott AG, Mainz [1]
E-mail: hartmut.hoeness@schott.com

Alfred Jacobsen
Schott AG, Mainz [1]
retired

Werner Kiefer
Schott AG, Mainz [1]
retired

Konrad Knapp
Schott Lithotec AG
Otto-Schott-Straße 13
07745 Jena

Ernst-Dieter Knohl
formerly Carl Zeiss,
73446 Oberkochen,
Germany

Ioannis Kosmas
Schott AG, Mainz [1]
E-mail: ioannis.kosmas@schott.com

[1] Hattenbergstraße 10, 55122 Mainz, Germany

Cora Krause
Schott AG, Mainz [1]
E-mail: cora.krause@schott.com

Klaus Kristen†

Roland Leroux
Schott AG, Mainz [1]
E-mail: roland.leroux@schott.com

Thomas Marx
Am Waldsaum 6,
58452 Witten-Bommern,
Germany,
formerly
Schott Glass Technologies,
400 York Avenue,
Duryea, PA 18642-2036,
USA

Hans Morian
Schott AG, Mainz [1]
retired

Gerd Müller
Fraunhofer-Institut
für Silicatforschung,
Neunerplatz 2,
97082 Würzburg,
Germany

Rudolf Müller
Schott AG, Mainz [1]
E-mail: rudolf.mueller@schott.com

Toni Münch
Schott AG, Mainz [1]
E-mail: toni.muench@schott.com

Peter Naß
Schott AG, Mainz [1]
E-mail: peter.nass@schott.com

Wolfgang Pannhorst
Schott AG, Mainz [1]
E-mail: wolfgang.pannhorst@schott.com

Norbert Reisert
Schott AG, Mainz [1]
E-mail: norbert.reisert@schott.com

Erich W. Rodek
Schott AG, Mainz [1]
E-mail: erich.rodek@schott.com

Kurt Schaupert
Schott AG, Mainz [1]
E-mail: kurt.schaupert@schott.com

Herwig Scheidler
Schott AG, Mainz [1]
E-mail: herwig.scheidler@schott.com

Ulrich Schiffner
Schott AG, Mainz [1]
E-mail: Ulrich Schiffner@schott.com

Hinnerk Schildt†

Wolfgang Schmidbauer
Schott AG, Mainz [1]
E-mail: wolfgang.schmidbauer@schott.com

Patrik Schober
Schott AG, Mainz [1]
E-mail: patrik.schober@schott.com

Fritz Schröder
Schott AG, Mainz [1]
E-mail: fritz.schroeder@schott.com

Fritz Siebers
Schott AG, Mainz [1]
E-mail: friedrich.siebers@schott.com

[1] Hattenbergstraße 10, 55122 Mainz, Germany

Burkhard Speit
Schott AG, Mainz [1]
E-mail: burkhard.speit@schott.com

Martin Taplan
Schott AG, Mainz [1]
E-mail: martin.taplan@schott.com

Armin Thomas
Schott AG, Mainz [1]
E-mail: armin.thomas@schott.com

Ted Wegert
Schott Home Tech North America,
5680 Shepherdsville Road,
Louisville, KY 40228,
USA
E-mail: tawegert@vol.com

Waldemar Weinberg
formerly Schott AG, Mainz [1]

Evelin Weiss
Schott AG, Mainz [1]

Dietmar Wennemann
Schott AG, Mainz [1]
E-mail: dietmar.wennemann
@schott.com

Eva Willhauk
Schott AG, Mainz [1]
E-mail: eva.willhauk@schott.com

[1] Hattenbergstraße 10, 55122 Mainz, Germany

Sources of Figures

We are indebted to the following editors and authors, respectively, for the kind permission to reproduce copyrighted materials.

Material	Source	Original Publisher
Fig. 2.20	[2.70]	The American Ceramic Society, 735 Ceramic Place, Westerville, Ohio 43081-8720, USA
Fig. 2.24	[2.62]	The American Ceramic Society, 735 Ceramic Place, Westerville, Ohio 43081-8720, USA
Fig. 2.25	[2.62]	The American Ceramic Society, 735 Ceramic Place, Westerville, Ohio 43081-8720, USA
Fig. 3.12	[3.3]	Rinnai Corporation 2–26, Fukuzumi, Nakagawa, Nagoya, Aichi 454, Japan
Fig. 3.27	[3.33]	Elsevier Science B.V., Amsterdam Publishing Division, P.O. Box 521, 1000 AM Amsterdam, The Netherlands
Fig. 3.33	[3.46]	Deutsche Glastechnische Gesellschaft, Mendelssohnstr. 75–77, 60325 Frankfurt, Germany
Fig. 3.39	[3.48]	BSH Bosch und Siemens Hausgeräte GmbH, Carl-Wery-Str. 34, 81739 München, Germany
Fig. 3.41	[3.49]	TGK Fachzeitschrift, Aumann KG, Wittelsbacher Str. 23, Postfach 1625, 95100 Selb, Germany
Fig. 3.42	[3.49]	TGK Fachzeitschrift, Aumann KG, Wittelsbacher Str. 23, Postfach 1625, 95100 Selb, Germany
Fig. 4.1	[4.7]	Deutsche Glastechnische Gesellschaft, Mendelssohnstr. 75–77, 60325 Frankfurt, Germany
Fig. 4.4	[4.12]	Deutsche Glastechnische Gesellschaft, Mendelssohnstr. 75–77, 60325 Frankfurt, Germany
Fig. 4.5	[4.12]	Deutsche Glastechnische Gesellschaft, Mendelssohnstr. 75–77, 60325 Frankfurt, Germany
Fig. 4.6		S. P. G. Sprechsaal Publishing Corporation, P.O. Box 2962, 96418 Coburg, Germany
Fig. 4.7		Elsevier Science B.V., Amsterdam Publishing Division, P.O. Box 521, 1000 AM Amsterdam, The Netherlands

Fig. 4.9	[4.47]	Optical Society of America, Editorial Dept., Applied Optics, 1613 19th St., N.W. Washington, D.C. 20009, USA

Index

Printing: Krips bv, Meppel
Binding: Stürtz, Würzburg